R... ision ...
Practice i...
Arithmetic

R. L. Bolt

Edward Arnold

First published 1980
by Edward Arnold (Publishers) Ltd.
41 Bedford Square, London WC1B 3DQ

Answers are published in a separate book.

Also available in this series
Revision and Practice in Mathematics by R. L. Bolt

British Library Cataloguing in Publication Data

Bolt, Ronald Leonard
Revision and practice in arithmetic.
1. Arithmetic–Problems, exercises, etc.
I. Title
513′.076 QA139

ISBN 0-7131-0428-7

Typeset by Reproduction Drawings Ltd., Sutton, Surrey.

Printed in Great Britain
by Spottiswoode Ballantyne Limited, Colchester and London

Contents

Units of measurement

Length

1 centimetre (cm)	= 10 millimetres (mm)
1 metre (m)	= 100 centimetres
1 kilometre (km)	= 1000 metres

Mass

1 kilogram (kg)	= 1000 grams (g)
1 tonne (t)	= 1000 kilograms

Area

1 square centimetre (cm^2)	= 100 square millimetres (mm^2)
1 square metre (m^2)	= 10 000 square centimetres
1 are (a)	= 100 square metres
1 hectare (ha)	= 100 ares
1 square kilometre (km^2)	= 100 hectares

Volume

1 litre (l)	= 1000 cubic centimetres (cm^3)
1 millilitre (ml)	= $\frac{1}{1000}$ litre = 1 cubic centimetre

Time

1 minute (min)	= 60 seconds (s)
1 hour (h)	= 60 minutes
1 day	= 24 hours
1 year	= 365 days (366 in a leap year)

Calculating Aids

Mathematical tables, slide rules and electronic calculators should not be used in Exercises 1 to 10 and their use is not recommended in parts of certain other exercises. Elsewhere they should not be used for simple calculations.

Number system, Base ten

The **sum** of 5 and 7 is $5 + 7 = 12$
The **difference** of 5 and 7 is $7 - 5 = 2$
The **product** of 5 and 7 is $5 \times 7 = 35$
$36 \div 9 = 4$ 4 is the **quotient** when 36 is divided by 9.
The **operators** $+, -, \times, \div$. Use $\times$ and $\div$ first and then $+$ and $-$.

Examples
$30 - 4 \times 7 = 30 - 28 = 2$; $18 \div 2 + 4 = 9 + 4 = 13$

Brackets
Work out the inside of brackets first.

$(30 - 4) \times 7 = 26 \times 7 = 182$; $18 \div (2 + 4) = 18 \div 6 = 3$

Exercise 1

1. In the number 5324, the 2 means two tens. What does the 3 mean? What does the 5 mean?
2. In the number 7960, the 7 means seven thousand. What does the 9 mean? What does the 6 mean?
3. Express in figures (a) six thousand and eight (b) three thousand and seventy (c) twelve thousand four hundred and six.
4. Write in words (a) 230 (b) 5006 (c) 2090.
5. State the answers to (a) $63 + 7$ (b) $82 - 15$ (c) $29 + 14$ (d) $112 - 77$.
6. State the answers to (a) $72 + 83$ (b) $8 + 72$ (c) $100 - 34$ (d) $246 - 182$.
7. State the answers to (a) $7 + 70 + 77$ (b) $500 - 48$ (c) $808 - 80$.
8. State the answers to (a) $505 + 55 + 5$ (b) $1637 - 984$ (c) $620 - 572$
9. State the answers to (a) 23×6 (b) 54×5 (c) 126×8
10. State the answers to (a) 37×4 (b) 62×7 (c) 86×9
11. State the answers to (a) $185 \div 5$ (b) $138 \div 6$ (c) $816 \div 8$
12. State the answers to (a) $721 \div 7$ (b) $144 \div 4$ (c) $2358 \div 9$

13. Copy and complete this table:

Numbers	7, 4	5, 9	6, 23	37, 10
Sum of numbers	11			
Difference of numbers		4		
Product of numbers	28			

14. State the quotient when (a) 176 is divided by 8 (b) when 765 is divided by 5.
15. Rearrange the figures in 4758 so that you get the largest possible number.
16. Rearrange the figures in 9016 so that you get the smallest possible number.
17. Copy the following and fill in the missing figures indicated by asterisks:

(a)
```
  34
+ **
  --
  88
```
(b)
```
  *9
+ 2*
  --
  74
```
(c)
```
  *3
- 3*
  --
  55
```
(d)
```
  6*2*
- *9*2
  ----
  2326
```

18. (a) Find the product of 2, 5 and 13.
(b) Which two numbers have a product of 10? (Do not use 1.)
(c) Which three numbers have a product of 70? (Do not use 1.)
19. State the next two numbers in each of the following sequences:
(a) 3, 5, 7, 9, 11, . . ., . . . (b) 1, 2, 4, 8, 16, . . ., . . .
(c) 1, 3, 6, 10, 15, . . ., . . . (d) 1, 4, 9, 16, 25, . . ., . . .
20. State the next two numbers in each of the following sequences:
(a) 2, 5, 8, 11, 14, . . ., . . . (b) 0, 3, 6, 9, 12, . . ., . . .
(c) 20, 17, 14, 11, 8, . . ., . . . (d) 3, 4, 6, 9, 13, . . ., . . .
21. Evaluate the given expressions:
(a) $(2 + 3) \times 7$ (b) $2 + (3 \times 7)$ (c) $(24 - 10) - 2$
(d) $24 - (10 - 2)$ (e) $48 \div (4 \times 3)$ (f) $(48 \div 4) \times 3$
22. Evaluate the given expressions:
(a) $5 + (2 \times 4)$ (b) $(5 + 2) \times 4$ (c) $18 - (8 - 4)$
(d) $(18 - 8) - 4$ (e) $40 \div (10 \div 2)$ (f) $(40 \div 10) \div 2$
23. Evaluate the following expressions: (Remember to use $\times$ and $\div$ first and then + and –.)
(a) $18 - 3 \times 2$ (b) $4 \times 5 + 6$ (c) $10 \div 2 + 3$
(d) $8 + 12 \div 4$ (e) $(4 + 6) \times 3 - 1$ (f) $16 - 3(7 - 5)$
24. Evaluate the following expressions:
(a) $4 \times 7 - 20$ (b) $6 + 5 \times 3$ (c) $12 \div 4 + 2$
(d) $20 - 10 \div 2$ (e) $2(9 - 4) - 6$ (f) $4 + 5(3 - 2)$
25. What must be added to 39 so that it can be divided exactly by 9?
26. What must be added to 22 so that it can be divided exactly by 5?
27.

2		
6		8

Copy this square and put a number in each of the empty places so that the sum of the numbers in each row, in each column and in each diagonal is 15.

28. Find the smallest number which can be divided exactly by 3, 4 and 5. When a packet of sweets is shared equally among 2, 3, 4 or 5 children there is always one sweet left over. State the smallest possible number of sweets in the packet.

Find the value of:

29. 27×13	30. 94×46	31. 82×57	32. 136×82
33. 305×93	34. 412×76	35. $713 \div 31$	36. $1768 \div 52$
37. $2204 \div 29$	38. $4263 \div 49$	39. $3591 \div 63$	40. $7866 \div 38$

41. The product of three numbers is 819. Two of them are 13 and 7. What is the other?
42. The product of three numbers is 9315. Two of them are 15 and 23. What is the other?
43. State the answer to (a) 7×3 (b) 7×30 (c) 70×30 (d) 700×30.
44. State the answer to (a) 9×8 (b) 90×8 (c) 90×80 (d) 90×800.
45. State the answer to (a) $54 \div 6$ (b) $540 \div 6$ (c) $540 \div 60$ (d) $5400 \div 60$.
46. State the answer to (a) $35 \div 5$ (b) $350 \div 5$ (c) $3500 \div 5$ (d) $3500 \div 50$.
47. State the answer to (a) 62×20 (b) 14×300 (c) $180 \div 30$ (d) $350 \div 7$
48. State the answer to (a) 23×30 (b) 46×50 (c) $420 \div 6$ (d) $400 \div 80$.
49. When 29 is divided by 6 the quotient is 4 and the remainder is 5. Find the quotient and remainder when (a) 17 is divided by 3 (b) 38 is divided by 5 (c) 67 is divided by 10.
50. Find the quotient and remainder when (a) 19 is divided by 4 (b) 25 is divided by 7 (c) 45 is divided by 5.
51. When a certain number is divided by 7 the quotient is 2 and the remainder is 3. Find the number.
52. When a certain number is divided by 10 the quotient is 3 and the remainder is 8. Find the number.
53. xy means $x \times y$. Thus if $x = 7$ and $y = 4$, $xy = 7 \times 4 = 28$. Find xy if (a) $x = 9, y = 5$ (b) $x = 20, y = 11$
54. Find rh if (a) $r = 8, h = 6$ (b) $r = 12, h = 27$

The number 10 can be divided exactly by 1, 2, 5 and 10. These are the **factors** of 10.
A number which has no factors except 1 and itself is called a **prime number**. The first five prime numbers are 2, 3, 5, 7 and 11. (We do not regard 1 as a prime number.)
Even numbers have 2 as a factor. They end in 0, 2, 4, 6 or 8.
Odd numbers do not have 2 as a factor.
The **multiples** of 7 are 7, 14, 21, 28, etc.
18 is a multiple of 2, of 3, of 6, of 9 and of 18.
Multiples of 5 end in 5 or 0.

Index form
$125 = 5 \times 5 \times 5 = 5^3$. The ^{3} is an *index*. (Plural: indices)

$$81 = 9 \times 9 = 3 \times 3 \times 3 \times 3 = 3^4$$

Example Find the value of 2^6
$2^6 = 2 \times 2 \times 2 \times 2 \times 2 \times 2 = 4 \times 2 \times 2 \times 2 \times 2$
$= 8 \times 2 \times 2 \times 2 = 16 \times 2 \times 2 = 32 \times 2 = 64$

Example Express 40 as the product of prime factors.

$$40 = 8 \times 5 = 2 \times 4 \times 5 = 2 \times 2 \times 2 \times 5$$

This answer could be expressed as $2^3 \times 5$

Exercise 2

1. Which of the following are even numbers: 23, 56, 74, 86, 99?
2. Which of the following are odd numbers: 18, 27, 34, 43, 79?
3. Which of the following are multiples of 3: 21, 36, 51, 76, 84?
4. Which of the following are multiples of 7: 42, 57, 77, 112, 128?
5. Which of the following are multiples of 5: 35, 54, 160, 552, 1740?
6. List the prime numbers which are less than 40.
7. List the prime numbers between 40 and 80.
8. The prime numbers 3 and 17 add up to 20. Which other two prime numbers add up to 20?
9. The prime numbers 3 and 37 add up to 40. What other pairs of prime numbers add up to 40?
10. The factors of 6 are 1, 2, 3 and 6. List the factors of (a) 10 (b) 12 (c) 22 (d) 31 (e) 36
11. List the factors of (a) 9 (b) 14 (c) 16 (d) 21 (e) 23
12. (a) Write down all pairs of odd numbers which give 10 when added together.
    (b) Write down all pairs of prime numbers which give 30 when added together.
13. Write in index form: $3 \times 3 \times 3 \times 3$ and $2 \times 2 \times 2 \times 2 \times 2$
14. Write in index form: $7 \times 7 \times 7$ and $13 \times 13 \times 13 \times 13 \times 13$
15. State the values of: $5^2, 2^3, 7^2, 2^5$
16. State the values of: $4^2, 10^2, 3^3, 1^4$
17. x^2 means $x \times x$ and x^3 means $x \times x \times x$. Thus if $x = 8$,
    $x^2 = x \times x = 8 \times 8 = 64$ and $x^3 = 8 \times 8 \times 8 = 64 \times 8 = 512$.
    Find the value of x^2 and of x^3 if $x = 4$.
18. Find the value of r^2 and of r^3 if $r = 6$.
19. Express as 3^n: 9, 27, and 81
20. Express as 10^n: 100, 10 000 and 100 000.
21. Work out the values of 7^3, 4^4 and 3^5.
22. Work out the values of $2^2 \times 3$, 2×3^2, $2^3 \times 5$ and $2^2 \times 5^2$

23. $28 = 4 \times 7 = 2 \times 2 \times 7 = 2^2 \times 7$. Express as the product of prime factors in this way: 18, 20, 30, 36, 42 and 45.
24. Express as the product of prime factors: 12, 24, 54, 32, 60, 315.
25. Write down (a) the multiples of 6 up to 96 (b) the multiples of 8 up to 96 (c) the common multiples of 6 and 8 up to 96 (d) the lowest common multiple of 6 and 8.
26. Repeat question **23** for the numbers 9 and 15 up to 135.
27. (a) List the factors of 24
(b) List the factors of 30
(c) State the common factors of 24 and 30
(d) Which is the highest common factor of 24 and 30?
28. Repeat question **25** for the numbers 16 and 24.

Other number systems

Base ten (denary)

The number 234_{ten} means 2 hundreds + 3 tens + 4 units which can be written
$2 \times 10^2 + 3 \times 10 + 4$

Base six

In base six we count 1, 2, 3, 4, 5, 10, 11, 12, 13, 14, 15, 20, 21, etc.
The number 45 means 4 sixes + 5 units = $24_{ten} + 5 = 29_{ten}$.
The number 234_{six} means 2 thirty-sixes + 3 sixes + 4 units which can be written
$2 \times 6^2 + 3 \times 6 + 4$.
Hence $234_{six} = 2 \times 36_{ten} + 3 \times 6 + 4$
$= 72_{ten} + 18_{ten} + 4 = 94_{ten}$
$53_{ten} = 36_{ten} + 17_{ten} = 36_{ten} + 12_{ten} + 5$
$= 36_{ten} + 2 \times 6 + 5 = 125_{six}$
A denary number such as 161_{ten} can be changed to a base six number by repeated division by 6:

6 | 161 units
6 | 26 sixes, remainder 5 units
4 thirty-sixes, remainder 2 sixes

$161_{ten} = 425_{six}$

Addition and subtraction in base six

$$2 + 4 = 10_{six},\quad 10_{six} - 1 = 5,$$
$$5 + 3 = 6 + 2 = 10_{six} + 2 = 12_{six}$$

$$\begin{array}{r} 125_{six} \\ +\ \ 43_{six} \\ \hline 212_{six} \\ \hline \end{array} \qquad \begin{array}{r} 420_{six} \\ -\ 131_{six} \\ \hline 245_{six} \\ \hline \end{array}$$

Binary (base two) numbers

10_{two} means two, 100_{two} means four, 1000_{two} means eight
We count 1, 10, 11, 100, 101, 110, 111, 1000, 1001, etc.
$10\,110_{two}$ means 1 sixteen + 0 eight + 1 four + 1 two + 0 unit which can be written
$1 \times 2^4 + 0 \times 2^3 + 1 \times 2^2 + 1 \times 2 + 0.$
Hence $10\,110_{two} = 16_{ten} + 4 + 2 = 22_{ten}$

Exercise 3

1. Express as base ten numbers: 10_{six} 12_{six}, 100_{six}, 103_{six}, 110_{six}
2. Express as base ten numbers: 30_{six}, 34_{six}, 200_{six}, 205_{six}, 220_{six}
3. Express as base six numbers: 12_{ten}, 16_{ten}, 24_{ten}, 29_{ten}, 30_{ten}.
4. Express as base six numbers: 18_{ten}, 20_{ten}, 36_{ten}, 42_{ten}, 44_{ten}.
5. Evaluate in base six: $5 + 1$, $4 + 5$, $12_{six} + 4$, $10_{six} - 2$, $10_{six} - 4$.
6. Evaluate in base six: $4 + 3$, $5 + 5$, $23_{six} + 3$, $20_{six} - 1$, $20_{six} - 12_{six}$.
7. Write out the first ten numbers in the binary system (base two).
8. State the binary numbers for the denary numbers 4, 8, 16 and 32.
9. What binary numbers correspond to the denary (base ten) numbers 5, 9, 10 and 12?
10. What binary numbers correspond to the denary numbers 11, 17, 18 and 21?
11. What denary numbers correspond to the binary numbers 11, 100, 110 and 111?
12. What denary numbers correspond to the binary numbers 101, 1000, 1011 and 11 000?

Evaluate in the binary system:

13. $11 + 1$, $110 + 101$, $100 - 1$ and $100 - 10$.
14. $101 + 101$, $1011 + 11$, $100 - 11$ and $1011 - 110$.
15. 11×10, 110×11 and 101×101.
16. 101×11, 1101×101 and 1011×110.
17. Arrange the following binary numbers in order of size with the smallest first: 1101, 10 001, 1011, 10 100.

18. Arrange the following binary numbers in order of size with the largest first: 1010, 1001, 1100, 10 000.
19. Write out the first ten numbers in base three.
20. State the base three numbers for the denary numbers 3, 9, 27 and 81.
21. Express as base ten numbers: 20_{three}, 102_{three}, 211_{three}.
22. Express as base ten numbers: 22_{three}, 120_{three}, 1110_{three}.
23. Express as base three numbers: 12_{ten}, 16_{ten}, 26_{ten}.
24. Express as base three numbers: 15_{ten}, 23_{ten}, 47_{ten}.

Evaluate in base three:

25. $1 + 2$, $2 + 2$, $22_{three} + 1$, $10_{three} - 1$ and $100_{three} - 2$
26. $12_{three} + 2$, $212_{three} + 22_{three}$, $120_{three} + 222_{three}$, $11_{three} - 2$ and $101_{three} - 22_{three}$.
27. $22_{three} \times 2$ and $12_{three} \times 21_{three}$.
28. $22_{three} \times 11_{three}$ and $212_{three} \times 21_{three}$.
29. State the value of the 3 in 136_{seven}. Also state the value of the 1.
30. State the value of the 4 in 241_{five}. Also state the value of the 2.
31. Express in base ten: 53_{seven}, 243_{seven} and 1000_{seven}.
32. Express in base ten: 14_{five}, 33_{five} and 212_{five}.
33. Express as base five numbers: 15_{ten}, 22_{ten} and 137_{ten}.
34. Express as base seven numbers: 20_{ten}, 49_{ten} and 234_{ten}.
35. Express in base ten: 25_{eight}, 100_{eight} and 307_{eight}.
36. Express as base eight numbers: 10_{ten}, 32_{ten} and 219_{ten}.

Evaluate the following in the given bases:

37. $32_{five} + 3$
38. $222_{five} + 303_{five}$
39. $10_{five} - 3$
40. $101_{five} + 4$
41. $253_{six} + 34_{six}$
42. $415_{seven} + 560_{seven}$
43. $42_{six} - 5$
44. $132_{seven} - 45_{seven}$
45. $14_{five} \times 3$
46. $23_{six} \times 4$
47. $44_{seven} \times 5$
48. $36_{eight} \times 2$
49. $13_{five} \times 2$
50. $22_{five} \times 3$
51. $202_{five} \times 14_{five}$
52. $142_{five} \times 24_{five}$
53. $213_{four} \times 32_{four}$
54. $142_{six} \times 32_{six}$

Fractions

$\frac{2}{3} = \frac{2 \times 5}{3 \times 5} = \frac{10}{15}$ $\frac{2}{3}$ and $\frac{10}{15}$ are **equivalent fractions.**

$\frac{36}{45} = \frac{36 \div 9}{45 \div 9} = \frac{4}{5}$ $\frac{36}{45}$ has been **reduced to its lowest terms.**

$3\frac{2}{5} = 3 + \frac{2}{5} = \frac{15}{5} + \frac{2}{5} = \frac{17}{5}$. This is an **improper fraction.**

$\frac{47}{6} = \frac{42}{6} + \frac{5}{6} = 7 + \frac{5}{6} = 7\frac{5}{6}$. This is a **mixed number.**

Addition of fractions

$$\frac{7}{9} + \frac{1}{3} + \frac{5}{6} = \frac{14}{18} + \frac{6}{18} + \frac{15}{18} = \frac{14 + 6 + 15}{18} = \frac{35}{18} = 1\frac{17}{18}$$

Subtraction of fractions

$$7\frac{1}{2} - 2\frac{4}{5} = 7 + \frac{5}{10} - 2 - \frac{8}{10} = 5 - \frac{3}{10} = 4\frac{7}{10}$$

Multiplication of fractions

$$\frac{3}{4} \times \frac{2}{5} = \frac{6}{20} = \frac{3}{10};\quad 3\frac{1}{3} \times 1\frac{3}{4} = \frac{10}{3} \times \frac{7}{4} = \frac{70}{12} = \frac{35}{6} = 5\frac{5}{6}$$

Division of fractions

$$4\frac{1}{5} \div 1\frac{1}{2} = \frac{21}{5} \div \frac{3}{2} = \frac{21}{5} \times \frac{2}{3} = \frac{42}{15} = \frac{14}{5} = 2\frac{4}{5}$$

$$2\frac{3}{4} \div 5 = \frac{11}{4} \div \frac{5}{1} = \frac{11}{4} \times \frac{1}{5} = \frac{11}{20}$$

Exercise 4

1. Copy and complete: $\frac{2}{3} = \frac{}{6}, \frac{3}{4} = \frac{}{12}, \frac{1}{5} = \frac{}{15}, 1 = \frac{}{7}$
2. Copy and complete: $\frac{2}{5} = \frac{}{10}, \frac{1}{3} = \frac{}{21}, \frac{5}{6} = \frac{}{18}, 1 = \frac{}{4}$
3. Copy and complete: $3 = \frac{}{4}, 5 = \frac{}{3}, 1\frac{2}{3} = \frac{}{3}, 2\frac{1}{2} = \frac{}{2}$
4. Copy and complete: $4 = \frac{}{5}, 5 = \frac{}{9}, 3\frac{1}{4} = \frac{}{4}, 3\frac{2}{5} = \frac{}{5}$
5. $\frac{23}{5} = 4\frac{3}{5}$, a mixed number. Express as mixed numbers: $\frac{7}{2}, \frac{11}{3}, \frac{9}{4}, \frac{33}{5}$.
6. Express as mixed numbers: $\frac{9}{2}, \frac{8}{3}, \frac{25}{6}, \frac{19}{7}$.
7. Express as twelfths: $\frac{1}{2}, \frac{1}{3}, \frac{1}{4}, \frac{1}{6}, \frac{3}{4}$ and $\frac{5}{6}$.
8. Express as twentieths: $\frac{1}{10}, \frac{1}{4}, \frac{1}{5}, \frac{2}{5}, \frac{3}{4}$ and $\frac{7}{10}$

Place in order of size with the smallest first:

9. $\frac{4}{9}, \frac{7}{9}, \frac{2}{9}$
10. $\frac{5}{7}, \frac{5}{11}, \frac{5}{9}$
11. $\frac{2}{5}, \frac{3}{4}, \frac{7}{10}$ (Use Question 8)
12. $\frac{5}{6}, \frac{3}{4}, \frac{2}{3}$ (First change each into twelfths.)
13. $\frac{1}{2}, \frac{3}{8}, \frac{5}{12}$
14. $\frac{4}{5}, \frac{5}{6}, \frac{7}{10}$

Reduce the following fractions to their lowest terms:

15. $\frac{4}{10}, \frac{8}{12}$ and $\frac{3}{15}$
16. $\frac{6}{9}, \frac{21}{28}$ and $\frac{6}{18}$
17. $\frac{6}{8}, \frac{10}{15}$ and $\frac{12}{20}$
18. $\frac{12}{21}, \frac{20}{25}$ and $\frac{14}{35}$

Simplify:

19. $\frac{1}{2} + \frac{1}{3}$
20. $\frac{1}{4} + \frac{2}{5}$
21. $\frac{2}{3} + \frac{3}{4}$
22. $\frac{3}{5} + \frac{3}{10}$
23. $\frac{7}{9} + \frac{5}{3}$
24. $\frac{5}{6} + \frac{3}{4}$
25. $\frac{1}{2} + \frac{1}{3} + \frac{1}{6}$
26. $\frac{3}{4} + \frac{1}{2} + \frac{7}{8}$

27. $\frac{2}{5} - \frac{1}{10}$ 28. $\frac{2}{3} - \frac{1}{4}$ 29. $\frac{3}{4} - \frac{5}{8}$ 30. $\frac{4}{5} - \frac{2}{7}$
31. $\frac{3}{4} + \frac{1}{6} - \frac{1}{2}$ 32. $\frac{1}{2} + \frac{2}{5} - \frac{3}{10}$ 33. $\frac{3}{8} + \frac{1}{2} - \frac{7}{8}$ 34. $\frac{2}{3} + \frac{2}{9} - \frac{5}{6}$
35. $1 - \frac{5}{7}$ 36. $2 - \frac{3}{4}$ 37. $5 - 1\frac{1}{3}$ 38. $6 - 4\frac{1}{4}$
39. $1\frac{1}{3} + 2\frac{1}{4}$ 40. $3\frac{2}{5} + 1\frac{3}{10}$ 41. $2\frac{3}{4} + \frac{7}{8}$ 42. $1\frac{5}{6} + 2\frac{2}{3}$
43. $3\frac{1}{2} - 1\frac{1}{4}$ 44. $2\frac{1}{8} - \frac{3}{8}$ 45. $3\frac{1}{3} - 1\frac{5}{6}$ 46. $4\frac{1}{2} - 1\frac{3}{5}$
47. $\frac{2}{5} \times \frac{3}{7}$ 48. $\frac{5}{9} \times \frac{3}{4}$ 49. $1\frac{2}{3} \times \frac{7}{10}$ 50. $2\frac{1}{4} \times 2\frac{2}{3}$
51. $2\frac{1}{3} \times 3\frac{1}{7}$ 52. $3 \times \frac{4}{5}$ 53. $\frac{3}{7} \times 5$ 54. $1\frac{1}{4} \times 7$
55. $12 \times 3\frac{1}{4}$ 56. $2\frac{2}{3} \times 6$ 57. $\frac{2}{3} \div \frac{5}{7}$ 58. $\frac{4}{3} \div \frac{1}{2}$
59. $\frac{6}{7} \div \frac{3}{4}$ 60. $\frac{2}{3} \div \frac{5}{6}$ 61. $\frac{3}{8} \div \frac{1}{2}$ 62. $\frac{5}{6} \div \frac{1}{4}$
63. $2\frac{1}{3} \div \frac{3}{4}$ 64. $1\frac{1}{4} \div \frac{5}{8}$ 65. $\frac{9}{10} \div 2\frac{1}{4}$ 66. $5\frac{1}{3} \div 1\frac{3}{5}$
67. $\frac{6}{7} \div 2$ 68. $3\frac{1}{3} \div 5$ 69. $\frac{5}{7} \div 2$ 70. $1\frac{1}{3} \div 5$
71. $(\frac{3}{4})^2$ 72. $(1\frac{1}{3})^2$ 73. $(2\frac{1}{2})^2$ 74. $(1\frac{3}{4})^2$
75. $(\frac{1}{2} - \frac{1}{3}) \times \frac{1}{4}$ 76. $(\frac{3}{4} - \frac{1}{5}) \times 2\frac{1}{2}$ 77. $\frac{2}{3} \div (\frac{1}{2} + \frac{1}{3})$
78. $\frac{1}{4} + \frac{1}{5} - (\frac{1}{4} - \frac{1}{5})$ 79. $3\frac{1}{2} - (\frac{1}{3} + \frac{1}{6})$
80. $\frac{2}{3} \times (\frac{1}{4} - \frac{1}{6})$

State the value of:

81. $\frac{1}{2}$ hour in minutes 82. $\frac{1}{5}$ cm in mm 83. £$\frac{1}{2}$ in pence
84. £$\frac{1}{10}$ in pence 85. $\frac{1}{4}$ minute in seconds 86. $\frac{3}{5}$ minute in seconds
87. $\frac{3}{4}$ hour in minutes 88. $\frac{5}{6}$ day in hours 89. £$\frac{3}{5}$ in pence
90. $\frac{2}{5}$ metre in centimetres

Express the first quantity as a fraction of the second and reduce the fraction to its lowest terms:

91. £15, £21 92. 35p, 40p 93. 80p, £1
94. 60 cm, 1 metre 95. 40 min, 1 hour
96. 45 seconds, 1 minute 97. 15 hours, 1 day
98. 4 hours, 1 day 99. £7, £21 100. £35, £56
101. £1.80, £4.50 102. £5.60, £7.20
103. Find the product of $\frac{1}{2}$ and $\frac{1}{3}$ 104. Find the sum of $\frac{1}{2}$ and $\frac{1}{3}$
105. Find the sum of $\frac{3}{8}$ and $\frac{1}{2}$ 106. Find the product of $\frac{3}{8}$ and $\frac{1}{2}$
107. Divide $\frac{2}{5}$ by $\frac{4}{25}$ 108. Find the product of $3\frac{1}{3}$ and $1\frac{1}{2}$
109. Insert a fraction between (a) $\frac{3}{7}$ and $\frac{5}{7}$ (b) $\frac{1}{4}$ and $\frac{5}{12}$
110. The fractions $\frac{1}{5}$, $\frac{x}{15}$ and $\frac{1}{3}$ are in order of size. What is x?
111. $\frac{3}{4}$ of a sum of money is £24. What is $\frac{1}{4}$ of the sum of money? What is the whole sum of money?
112. $\frac{5}{7}$ of a length is 70 cm. What is $\frac{1}{7}$ of the length? What is the whole length?
113. From a man's weekly wage, $\frac{1}{3}$ is deducted for tax and $\frac{1}{9}$ for insurance. What fraction remains? He gives $\frac{3}{5}$ of this to his wife. What fraction now remains? He now has £24. What was his wage?

114. During an influenza epidemic, $\frac{3}{5}$ of a class were absent. There were 10 pupils present. How many were there in the class?

115. Write down all proper fractions with denominators less than 5 and place them in order of size with the smallest first. (You should have five fractions.)

Decimals: basic operations

$\frac{3}{10} = 0.3$, $\frac{9}{100} = 0.09$, $\frac{7}{1000} = 0.007$, $\frac{27}{1000} = 0.027$

$\frac{9}{25} = \frac{9 \times 4}{25 \times 4} = \frac{36}{100} = 0.36$ $\quad 0.085 = \frac{85}{1000} = \frac{17}{200}$

$0.06 \times 0.7 = 0.042$ (2 dec. pl. + 1 dec. pl. in the numbers gives 3 dec. pl. in the answer.)

$2.12 \times 0.004 = 0.008\,48$ (2 dec. pl. + 3 dec. pl. in the numbers gives 5 dec. pl. in the answer.)

Number	0.3	0.05	8.4	0.0077
Number × 10	3	0.5	84	0.077
Number × 100	30	5	840	0.77

Number	0.8	6.9	0.033	452
Number ÷ 10	0.08	0.69	0.0033	45.2
Number ÷ 100	0.008	0.069	0.000 33	4.52

$5.68 \div 4000 = (5.68 \div 4) \div 1000 = 1.42 \div 1000 = 0.001\,42$

$$2.73 \div 0.7 = \frac{2.73}{0.7} = \frac{2.73 \times 10}{0.7 \times 10} = \frac{27.3}{7} = 3.9$$

Exercise 5

1. Write as decimals: $\frac{7}{10}$, $\frac{3}{100}$, $\frac{9}{1000}$, $\frac{47}{100}$, $\frac{23}{10}$.
2. Write as decimals: $\frac{21}{100}$, $\frac{17}{1000}$, $\frac{9}{100}$, $\frac{413}{100}$, $5\frac{1}{10}$.
3. Write as fractions: 0.1, 0.07, 0.003, 0.53, 8.79.
4. Write as fractions: 0.01, 0.029, 0.17, 3.9, 10.03.
5. Write as decimals: $\frac{1}{5}$, $\frac{1}{2}$, $\frac{4}{5}$, $\frac{1}{20}$, $\frac{9}{50}$.
6. Write as decimals: $\frac{3}{5}$, $\frac{13}{20}$, $\frac{47}{50}$, $\frac{1}{25}$, $\frac{13}{25}$.

7. Write as fractions and reduce to their lowest terms: 0.2, 0.4, 0.05, 0.35, 0.08, 0.76
8. Write as fractions and reduce to their lowest terms: 0.8, 0.95, 0.75, 0.04, 0.18, 0.64
9. Express as decimals: $\frac{3}{500}$, $\frac{7}{200}$, $\frac{161}{500}$, $\frac{1}{200}$
10. Express as decimals: $\frac{11}{500}$, $\frac{43}{200}$, $\frac{7}{250}$, $\frac{31}{125}$
11. Write as fractions and reduce to their lowest terms: 0.002, 0.015, 0.666, 0.024.
12. Write as fractions and reduce to their lowest terms: 0.005, 0.092, 0.088, 0.905.
13. In 0.637 the 6 means six tenths or $\frac{6}{10}$. Write in two such ways the meaning of the 3. Also write in two ways the meaning of the 7.
14. Write in two ways (as in question **13**) the meaning of the 4 in 0.409. Also write in two ways the meaning of the 9.

Evaluate the following:

15. $4.4 + 3.9$ **16.** $7.3 + 2.8$ **17.** $6.4 + 1.26$ **18.** $3.55 + 2.7$
19. $8.2 - 5.9$ **20.** $6.4 - 2.6$ **21.** $3.4 - 1.27$ **22.** $0.8 - 0.36$
23. $2.56 - 1.8$ **24.** $0.73 - 0.042$ **25.** $6.3 + 7.05 - 11.62$
26. $4.14 + 3.7 - 0.95$ **27.** $4.8 - 1.5 + 2.9$ **28.** $3.3 - 0.33 + 3.03$
29. 0.4×2 **30.** 0.8×3 **31.** 0.07×4 **32.** 0.17×8
33. 0.03×100 **34.** 0.7×100 **35.** 8.4×1000 **36.** 0.39×1000
37. 0.024×20 **38.** 3.3×200 **39.** 6.57×300 **40.** 2.45×200
41. 1.85×4000 **42.** 23.4×3000 **43.** $4.4 \div 10$ **44.** $0.8 \div 10$
45. $0.55 \div 100$ **46.** $8 \div 100$ **47.** $2.6 \div 1000$ **48.** $0.09 \div 100$
49. $3.5 \div 5$ **50.** $0.18 \div 3$ **51.** $0.036 \div 4$ **52.** $0.84 \div 7$
53. $0.7 \div 2$ **54.** $0.13 \div 2$ **55.** $0.9 \div 4$ **56.** $0.26 \div 4$
57. $1.4 \div 20$ **58.** $0.36 \div 30$ **59.** $4.5 \div 500$ **60.** $5.4 \div 600$
61. 0.4×0.2 **62.** 0.7×0.2 **63.** 0.02×0.8 **64.** 0.3×0.01
65. 6×0.02 **66.** 0.3×2.5 **67.** 0.5×0.06 **68.** 0.005×0.8
69. $0.3 \times 0.2 \times 0.8$ **70.** $0.9 \times 0.5 \times 0.4$
71. 2.3×1.7 **72.** 0.56×3.3 **73.** 27.7×1.4 **74.** 3.8×2.9
75. $0.06 \div 0.2$ **76.** $1.2 \div 0.03$ **77.** $5.5 \div 0.11$ **78.** $0.07 \div 0.02$
79. $1.5 \div 0.04$ **80.** $0.056 \div 0.8$ **81.** $1.488 \div 0.6$ **82.** $0.392 \div 0.07$
83. $16.28 \div 4.4$ **84.** $7.412 \div 1.7$ **85.** $3.528 \div 2.52$ **86.** $8.008 \div 2.6$
87. $3.536 \div 0.17$ **88.** $2.4157 \div 2.03$

89. Express 37 400 metres in kilometres.
90. Express 8040 grams in kilograms.
91. Express 0.06 kilometres in metres.
92. Express 0.57 kilograms in grams.

Decimals: approximations

Decimal places

Number	Correct to 3 dec. pl.	Correct to 2 dec. pl.	Correct to 1 dec. pl.
3.6924	3.692	3.69	3.7
0.4958	0.496	0.50	0.5

Significant figures

Number	Correct to 3 sig. fig.	Correct to 2 sig. fig.	Correct to 1 sig. fig.
8369	8370	8400	8000
5.026	5.03	5.0	5

$\frac{3}{7} = 3 \div 7 = 0.42857\ldots = 0.429$, correct to 3 dec. pl.

Approximate values

Example 1: By rounding each number to 1 sig. fig., find the value of 7.9 × 0.038, correct to 1 sig. fig.

7.9 is nearly 8 and 0.038 is nearly 0.04

$7.9 \times 0.038 \simeq 8 \times 0.04 = 0.32 \simeq 0.3$

Example 2: Find the value of $0.621 \div 0.0192$ correct to 1 sig. fig.

$$\frac{0.621}{0.0192} \simeq \frac{0.6}{0.02} = \frac{60}{2} = 30$$

Exercise 6

1. Express to the nearest metre:
   7.8 m, 5.4 m, 3.68 m, 24.48 m, 39.7 m.
2. Express to the nearest hour:
   3 h 40 min, 6 h 55 min, 7 h 25 min, 13 h 33 min.
3. Express to the nearest 10 kilometres:
   363 km, 528 km, 3208 km, 1674 km, 37 km.
4. Express to the nearest 10 grams:
   647 g, 214 g, 76 g, 2382 g, 7407 g.
5. Express to the nearest tenth of a second:
   7.46 s, 2.33 s, 0.084 s, 0.477 s, 3.089 s.

6. Express correct to one decimal place:
4.58, 8.24, 10.63, 35.472, 11.96.
7. Express correct to 2 decimal places:
3.643, 0.817, 0.0215, 0.9374, 8.098.
8. Express correct to 3 decimal places:
2.5792, 0.6438, 0.0517, 0.0042, 0.009 63.
9. Express correct to 2 significant figures:
517, 3.24, 0.643, 0.001 77, 3216.
10. Express correct to 3 significant figures:
2.324, 0.063 76, 25.02, 13 209, 146.54.
11. Express the number 3.141 593 correct to
(a) 5 sig. fig. (b) 4 sig. fig. (c) 2 sig. fig. (d) 2 dec. pl.
12. Express the number 0.072 68 correct to
(a) 3 dec. pl. (b) 3 sig. fig. (c) 2 sig. fig. (d) 2 dec. pl.

Calculate the following, giving each answer correct to 2 dec. pl.

13. $5 \div 7$ **14.** $0.83 \div 3$ **15.** $5.27 \div 6$ **16.** $0.1972 \div 0.67$

Calculate the following, giving each answer correct to 3 sig. fig.

17. $4 \div 9$ **18.** $44 \div 7$ **19.** $323 \div 6$ **20.** $0.1354 \div 3$

Express the following fractions as decimals, correct to 3 dec. pl.

21. $\frac{2}{7}$ **22.** $\frac{2}{3}$ **23.** $\frac{7}{9}$ **24.** $\frac{3}{11}$ **25.** $\frac{11}{15}$ **26.** $\frac{2}{13}$

Calculate, correct to 3 sig. fig., the value of:

27. 17.9×6.3 **28.** 32.6×1.3 **29.** $0.67 \div 4.8$ **30.** $55.8 \div 0.23$

Round each number to 1 sig. fig. and then find an approximate value, correct to 1 sig. fig. for the product or quotient:

31. 2.2×2.9 **32.** 1.9×4.3 **33.** 8.1×9.8
34. 97×2.2 **35.** 6.7×4.3 **36.** 4.2×5.82
37. 0.69×0.33 **38.** 0.073×1.8 **39.** 4.3×0.0362
40. 0.69×8.83 **41.** $5.9 \div 2.1$ **42.** $8.3 \div 1.9$
43. $19.2 \div 5.1$ **44.** $18.2 \div 3.9$ **45.** $0.83 \div 0.43$
46. $0.59 \div 0.19$ **47.** $0.079 \div 0.22$ **48.** $0.887 \div 0.032$

49. Which of the following is nearest to the value of 23.26×4.77:
10, 80, 100, 800 or 1000?
50. Which of the following is nearest to the value of 0.792×51.8:
0.04, 0.4, 4, 40, 400 or 4000?
51. Which of the following is nearest to the value of $6.42 \div 0.034$:
0.02, 0.2, 2, 20, 200 or 2000?
52. Which of the following is nearest to the value of $0.913 \div 29.7$:
0.03, 0.3, 3, 30, 300 or 3000?
53. Which of the following is nearest to the value of $\dfrac{0.0713 \times 6.14}{0.742}$
0.06, 0.6, 6, 60, 600 or 6000?
54. Which of the following is nearest to the value of $\dfrac{0.0294 \times 0.0807}{0.005\,89}$
0.04, 0.4, 4, 40, 400 or 4000?
55. Find, correct to 1 sig. fig., the value of (a) £6.85 × 49 (b) £186 × 32.

Money and measures

Exercise 7

1. Find the total cost of five coffees at 15p each and five cakes at 11p each.
2. A collecting box on a flag day contained the following coins: 42 of 1p, 64 of 2p, 95 of 5p and 26 of 10p. Find the total value.
3. I buy three 10p stamps and four 12p stamps. How much change do I get from a £1 note?
4. A housewife buys groceries costing 32p, 59p, 28p, 43p and 64p. How much change does she get from a £5 note?
5. The hourly rate for a job is 152p. What is the wage for a 40 hour week?
6. What is the total cost of 5 oranges at 7p each, 2 grapefruit at 12p each and 2 kg of potatoes at 18p per kg?
7. At a school play, 62 tickets at 45p and 94 tickets at 30p were sold. How much money was received?
8. A club has £42.15. It spends £33.82. How much is left?
9. A pupil has £72.40 in his savings account. He withdraws £58.90 to buy a bicycle. How much is left in the account?
10. Twenty pupils each pay £4 for an outing. The total cost of the outing is £76.40. (a) How much money is left over? (b) How much can be given back to each pupil?
11. A penny is 2 cm wide. A line of pennies is a metre long. What is the value of the coins?
12. How many 50 g bars of chocolate can be made from 15 kg?
13. A clock loses 50 seconds each day. After how many days is it 5 minutes slow?
14. A bucket weighs 7.6 kg when full of water and 4.4 kg when half full. What does it weigh when it is a quarter full?
15. (a) Express 6 cm in millimetres (b) Express 3 m in centimetres (c) Express 5 km in metres (d) Express 7 m in millimetres
16. (a) Express 400 mm in centimetres (b) Express 400 cm in metres (c) Express 9000 m in kilometres
17. (a) Express 8000 g in kilograms (b) Express 6000 kg in tonnes
18. (a) Express 7 hours in minutes (b) Express 3 h 35 min in minutes (c) Express 405 minutes in hours and minutes (d) Express 136 seconds in minutes and seconds
19. Express in metres: (a) 500 cm (b) 540 cm (c) 547 cm
20. Express in centimetres: (a) 6.3 m (b) 9.48 m (c) 17.7 m
21. Express in kilometres: (a) 8000 m (b) 8360 m (c) 45 m
22. Express in metres: (a) 0.9 km (b) 7.4 km (c) 3.25 km
23. Express in kilograms: (a) 4000 g (b) 300 g (c) 5600 g
24. The mass of a tin of fruit is 320 g. Find the mass in kilograms of 50 such tins.

25. A bag of coal has a mass of 50 kg. Find the mass in tonnes of 80 such bags.
26. An aircraft is carrying 80 passengers. The average mass of a passenger is 85 kg and the average mass of each passenger's luggage is 40 kg. What is the total mass in tonnes of the passengers and luggage?
27. A job is advertised at a starting salary of £2640 p.a. rising by annual increments of £160 to a maximum of £4560 p.a.
(a) Find the difference between the starting and maximum salaries.
(b) How many increments are there?
(c) What is the salary for the sixth year?

Indices and standard form

Indices

$8^3 \times 8^2 = (8 \times 8 \times 8) \times (8 \times 8) = 8 \times 8 \times 8 \times 8 \times 8 = 8^5$
Hence $8^3 \times 8^2 = 8^{3+2} = 8^5$
Similarly, $7^4 \times 7^6 = 7^{4+6} = 7^{10}$

$$10^6 \div 10^4 = \frac{10 \times 10 \times 10 \times 10 \times 10 \times 10}{10 \times 10 \times 10 \times 10} = 10 \times 10 = 10^2$$

Hence $10^6 \div 10^4 = 10^{6-4} = 10^2$
Similarly, $5^7 \div 5^3 = 5^{7-3} = 5^4$

3^{-2} means $\frac{1}{3^2} = \frac{1}{9}$, 10^{-3} means $\frac{1}{10^3} = \frac{1}{1000} = 0.001$

Adding and subtracting directed numbers
$(-3) + (-8) = (-11)$, $(-3) + 8 = 5$, $3 + (-8) = -5$
$(-3) - (-8) = -3 + 8 = 5$, $(-3) - 8 = -3 - 8 = -11$
$3 - (-8) = 3 + 8 = 11$, $3 - 8 = -5$

Exercise 8

1. Write in index form: $3 \times 3 \times 3 \times 3$, $5 \times 5 \times 5$, $10 \times 10 \times 10 \times 10 \times 10$
2. Express as 10^n: 1000, 1 million and one hundred million
3. Find the value of 2^3, 5^2, 3^3, 2^4 and 10^4
4. Find the value of 3^2, 2^5, 4^3, 5^3 and 10^5

5. Express as 2^n: $2^3 \times 2^4$, $2^5 \times 2$, $2^6 \times 2^4$
6. Express as 10^n: $10^2 \times 10^6$, $10^3 \times 10^3$, 10×10^8
7. Express as 3^n: $3^5 \div 3^2$, $3^{10} \div 3^4$, $3^7 \div 3$
8. Express as 10^n: $10^7 \div 10^3$, $10^4 \div 10$, $10^9 \div 10^8$
9. Express as fractions without indices: 3^{-1}, 3^{-2}, 3^{-3}
10. Express as fractions without indices and then as decimals: 10^{-1}, 10^{-2}, 10^{-4}
11. Find the value of: $(-7) + (-2)$, $7 + (-2)$, $(-7) + 2$
12. Find the value of: $(-10) + (-4)$, $(-10) + 4$, $10 + (-4)$
13. Find the value of: $(-2) - (-6)$, $(-2) - 6$, $2 - (-6)$, $2 - 6$
14. Find the value of: $(-9) - (-4)$, $(-9) - 4$, $9 - (-4)$, $9 - 4$
15. Find the value of: $(-2) - (-2)$, $(-2) - 2$, $2 - (-2)$, $2 - 2$
16. Find the value of: $5 - 11$, $5 - (-11)$, $(-5) - (-11)$, $(-5) - 11$
17. Express as 5^n; $5^{-2} \times 5^{-4}$, $5^{-1} \times 5^6$, $5^{-4} \times 5^2$
18. Express as 3^n: $3^{-3} \times 3^{-1}$, $3^7 \times 3^{-2}$, $3^3 \times 3^{-5}$
19. Express as 10^n: $10^{-3} \times 10^5$, $10^2 \times 10^{-7}$, $10^{-3} \times 10^{-4}$
20. Express as 2^n: $2^2 \div 2^5$, $2^2 \div 2^{-5}$, $2^{-2} \div 2^5$
21. Express as 10^y: $10^3 \div 10^7$, $10^{-3} \div 10^7$, $10^3 \div 10^{-7}$
22. Express in the form 10^n: $\frac{1}{100}$, 0.1, 0.000 01
23. Express in the form 10^n: $10^2 \times 10^3 \times 10^4$ and $10 \times 10^4 \times 10^{-3}$
24. Express in the form 10^n: $10^5 \times 10 \div 10^4$ and $10^2 \times 10^3 \div 10^7$

Standard form

Any positive number can be expressed in the form $a \times 10^n$ where $1 \leqslant a < 10$ and n is a positive or negative whole number.

Examples: $27\,300 = 2.73 \times 10\,000 = 2.73 \times 10^4$

$$0.064 = \frac{64}{1000} = \frac{6.4}{100} = 6.4 \times \frac{1}{100} = 6.4 \times 10^{-2}$$

Combining numbers in standard form

$$(7 \times 10^{-3}) + (2 \times 10^{-3}) = 9 \times 10^{-3}$$

$$(6 \times 10^5) \times (9 \times 10^3) = 54 \times 10^8 = 5.4 \times 10 \times 10^8 = 5.4 \times 10^9$$

$$(4.3 \times 10^2) \times (5 \times 10^{-5}) = 21.5 \times 10^{-3} = 2.15 \times 10 \times 10^{-3} = 2.15 \times 10^{-2}$$

$$(3 \times 10^3) \div (5 \times 10^7) = \frac{3 \times 10^3}{5 \times 10^7} = 0.6 \times 10^{-4} = 6 \times 10^{-1} \times 10^{-4} = 6 \times 10^{-5}$$

Exercise 9

Write the following numbers in standard form:

1. 270 2. 36 000 3. 4000 4. 308 000 5. 62.9
6. 3920 7. 1005 8. 70 000 9. 8 thousand
10. 3 million

Write as ordinary numbers, that is, without powers of 10:

11. 4.7×10^2 12. 5.6×10^4 13. 9×10^3 14. 1.342×10^2
15. 6.8×10^5

Write the following numbers in standard form:

16. 0.038 17. 0.0075 18. 0.926 19. 0.05 20. 0.008 07
21. 0.13 22. 0.0244 23. 0.309 24. 8 thousandths
25. 7 millionths

Write as ordinary numbers, that is, without powers of 10:

26. 8.2×10^{-2} 27. 3.4×10^{-4} 28. 3×10^{-3}
29. 2.75×10^{-1} 30. 7.8×10^{-3}

31. Express in standard form:
    (a) the distance of Mars from the Sun, 228 million kilometres
    (b) the speed of light, 300 000 kilometres per second
    (c) the wavelength of green light, 0.000 054 6 cm
32. Explain why the following are not in standard form:
    42×10^5, 0.37×10^4, $5.4 \div 10^6$, $7.9 \times 10^{0.3}$

$$360 \times 10^4 = 3.6 \times 10^2 \times 10^4 = 3.6 \times 10^6$$

Use this method to express the following numbers in standard form:

33. 58×10^3 34. 39×10^5 35. 940×10 36. $42\,000 \times 10^2$

$$0.037 \times 10^5 = 0.037 \times 100 \times 10^3 = 3.7 \times 10^3$$

Use this method to express the following in standard form:

37. 0.82×10^5 38. 0.6×10^7 39. 0.0053×10^5 40. 0.09×10^3

Work out the following, giving each answer in standard form:

41. $(3 \times 10^5) + (4 \times 10^5)$
42. $(6.5 \times 10^4) + (9.2 \times 10^4)$
43. $(7.6 \times 10^3) - (2.2 \times 10^3)$
44. $(8.5 \times 10^2) - (1.7 \times 10^2)$
45. $(3 \times 10^2) \times (2 \times 10^3)$
46. $(2.2 \times 10^4) \times (4 \times 10^3)$
47. $(6 \times 10^2) \times (7 \times 10^4)$
48. $(5.5 \times 10) \times (3 \times 10^2)$
49. $(4 \times 10^2) \times (3.5 \times 10^3)$
50. $(5 \times 10^5) \times (4.4 \times 10^7)$
51. $(8 \times 10^5) \div (2 \times 10^2)$
52. $(6 \times 10^7) \div (3 \times 10^5)$
53. $(7.2 \times 10^3) \div (4 \times 10)$
54. $(9.6 \times 10^{10}) \div (3 \times 10^7)$
55. $(2.4 \times 10^8) \div (4 \times 10^3)$
56. $(2 \times 10^6) \div (5 \times 10^2)$
57. $(4 \times 10^3) \times (2 \times 10^{-1})$
58. $(3 \times 10^2) \times (2 \times 10^{-5})$
59. $(3.6 \times 10^{-3}) \div (2 \times 10^{-1})$
60. $(9.6 \times 10^{-2}) \div (6 \times 10^{-5})$
61. $(4 \times 10^{-3}) \times (7 \times 10^6)$
62. $(9 \times 10^{-5}) \times (4 \times 10^2)$
63. $(1.54 \times 10^{-5}) \div (7 \times 10^2)$
64. $(2.7 \times 10^2) \div (9 \times 10^{-1})$

Powers and roots

$3^2 = 9$, $8^2 = 64$, $29^2 = 841$, $163^2 = 26\,569$
9, 64, 841 and 26 569 are **perfect squares.**
3, 8, 29 and 163 are their **square roots.**
The symbol $\sqrt{}$ is used for 'square root'. Thus $\sqrt{841} = 29$.

Square roots of perfect squares by factorisation
For $\sqrt{n}$ we seek a number r such that $n = r \times r$.
$225 = 3 \times 3 \times 5 \times 5 = (3 \times 5) \times (3 \times 5) = 15 \times 15$ and so $\sqrt{225} = 15$.
$784 = 4 \times 196 = 4 \times 4 \times 49 = 2 \times 2 \times 2 \times 2 \times 7 \times 7$
$= (2 \times 2 \times 7) \times (2 \times 2 \times 7) = 28 \times 28$
and so $\sqrt{784} = 28$.
$250\,000 = 25 \times 10\,000 = 5 \times 5 \times 100 \times 100 = (5 \times 100) \times (5 \times 100)$
$= 500 \times 500$
and so $\sqrt{250\,000} = 500$
$5^3 = 125$ and $8^3 = 512$. 125 and 512 are **perfect cubes.**
5 and 8 are their **cube roots.** $\sqrt[3]{125} = 5$ and $\sqrt[3]{512} = 8$.

Exercise 10

1. Write down the squares of 3, 5, 6 and 8.
2. Write down the squares of 10, 100 and 1000.
3. Write down the values of 1^2, 4^2, 9^2 and 11^2.
4. Evaluate 20^2, 70^2, 900^2 and 170^2.
5. State the square roots of 9, 16, 64 and 81.
6. State the square roots of 25, 49, 100 and 1.
7. Write down the values of $\sqrt{36}$, $\sqrt{100}$ and $\sqrt{4900}$.
8. Write down the values of $\sqrt{25}$, $\sqrt{900}$ and $\sqrt{3600}$.
9. 50 is the square root of which of the following: 250, 2500 or 25 000?
10. 20 is the square root of which of the following: 40, 400 or 4000?

Express each number as the product of prime factors and then find its square root:

11. 225 **12.** 196 **13.** 441 **14.** 324 **15.** 484 **16.** 625
17. 729 **18.** 784 **19.** 1225 **20.** 1089

21. Evaluate the square roots of 6400 and 640 000.
22. Evaluate the square roots of 40 000 and 9 000 000.
23. Find x in each of the following cases: (a) $x^2 = 49$ (b) $x^2 = 4^2 + 3^2$ (c) $x^2 = 10^2 - 8^2$
24. Find y in each of the following cases:
    (a) $y^2 = 121$ (b) $y^2 = 5^2 + 12^2$ (c) $y^2 = 17^2 - 15^2$
25. Write down the cubes of 2, 3, 4 and 10.
26. State the cube roots of 1, 8, 27, 64 and 1000.

27. Evaluate 5^3, 7^3 and 14^3.

28. State the values of $\sqrt[3]{125}$, $\sqrt[3]{343}$ and $\sqrt[3]{1}$.

Express each number as the product of prime factors and then find its cube root:

29. 216 30. 729 31. 3375 32. 2744

Decimals and fractions

$0.49 = 0.7 \times 0.7$ and so $\sqrt{0.49} = 0.7$

$0.0169 = 0.13 \times 0.13$ and so $\sqrt{0.0169} = 0.13$

$$\left(1\tfrac{2}{3}\right)^2 = \left(\tfrac{5}{3}\right)^2 = \tfrac{25}{9} = 2\tfrac{7}{9} \qquad \sqrt{5\tfrac{1}{16}} = \sqrt{\tfrac{81}{16}} = \frac{\sqrt{81}}{\sqrt{16}} = \tfrac{9}{4} = 2\tfrac{1}{4}$$

$$\sqrt{0.0036} = \sqrt{\frac{36}{10\,000}} = \frac{\sqrt{36}}{\sqrt{10\,000}} = \frac{6}{100} = 0.06$$

Exercise 11

Do not use tables or a calculator in Questions 1 to 20.

1. Calculate the squares of 0.3, 0.5, 0.03 and 0.05.
2. Calculate the squares of 0.2, 0.9, 0.01 and 0.07.
3. Find the square roots of 0.36, 0.09, 0.0004 and 0.0025.
4. Find the square roots of 0.01, 0.64, 0.0081 and 0.0009.
5. Find $\sqrt{0.0049}$, $\sqrt{0.49}$, $\sqrt{0.04}$ and $\sqrt{0.0004}$.
6. Find $\sqrt{0.09}$, $\sqrt{0.81}$, $\sqrt{0.0001}$ and $\sqrt{0.0016}$.
7. Calculate the squares of $1\frac{1}{2}$, $3\frac{1}{2}$ and $2\frac{2}{3}$.
8. Calculate the squares of $2\frac{1}{2}$, $1\frac{2}{3}$ and $3\frac{1}{3}$.
9. Calculate the square roots of $2\frac{1}{4}$, $7\frac{1}{9}$ and $5\frac{4}{9}$.
10. Calculate the square roots of $1\frac{7}{9}$, $6\frac{1}{4}$ and $1\frac{9}{16}$.
11. Calculate $\sqrt{225}$ and state the value of $\sqrt{2.25}$
12. Calculate $\sqrt{441}$ and state the value of $\sqrt{4.41}$.
13. Calculate $\sqrt{196}$ and state the values of $\sqrt{0.0196}$ and $\sqrt{0.000\,196}$.
14. Calculate $\sqrt{1225}$ and state the values of $\sqrt{0.1225}$ and $\sqrt{0.001\,225}$.
15. 0.06 is the square root of which of the following numbers: 0.036, 0.0036, 0.000 36?
16. 0.03 is the square root of which of the following: 0.09, 0.009, 0.0009?
17. Calculate the cubes of 0.2, 0.3, 0.4 and 0.04.
18. Calculate the cubes of 0.1, 0.5, 0.01 and 0.05.
19. Which of the following are perfect cubes: 0.8, 0.08, 0.008, 0.125, 0.0125?
20. Find the cube roots of 216, 0.216 and 0.000 216.

Use tables or a calculator to find the values of the following, correct to 3 sig. fig.:

21. 37.2^2 22. 168^2 23. 0.749^2 24. 0.108^2

25. $6.8^2 + 7.7^2$ 26. $0.95^2 + 0.87^2$

27. $9.64^2 - 5.03^2$ 28. $1.46^2 - 0.77^2$

Approximate square roots
29 is between 25 and 36 and so $\sqrt{29}$ is between 5 and 6.
$5.3^2 = 28.09$ and $5.4^2 = 29.16$ and so $\sqrt{29}$ is a little less than 5.4.
Tables or a calculator can be used for approximate square roots.
From a set of 4 figure tables $\sqrt{29} \simeq 5.385$.
From a calculator $\sqrt{29} \simeq 5.385\,164\,807$

Example: Use tables for $\sqrt{73\,290}$

$$73\,290 = 7.3290 \times 10\,000$$
$$\sqrt{73\,290} = \sqrt{7.3290} \times \sqrt{10\,000} \simeq 2.707 \times 100 = 270.7$$

Example: Use tables for $\sqrt{0.003\,18}$

$$0.003\,18 = 31.8 \div 10\,000$$
$$\sqrt{0.003\,18} = \sqrt{31.8} \div \sqrt{10\,000} \simeq 5.639 \div 100 = 0.056\,39$$

Exercise 12

Do not use tables or a calculator in Questions 1 to 7.

1. State the nearest integer to $\sqrt{18}, \sqrt{39}, \sqrt{68}, \sqrt{105}, \sqrt{79}$.

2. $\sqrt{55}$ is between 7 and 8. Make a statement like this for each of the following: $\sqrt{11}, \sqrt{22}, \sqrt{90}, \sqrt{114}$.

3. Calculate 3.4^2, 3.5^2, 3.6^2 and 3.7^2. Which of the numbers 3.4, 3.5, 3.6, 3.7 is nearest to $\sqrt{13}$?

4. Which of the following is nearest to $\sqrt{840}$: 9, 30, 90 or 300?

5. Which of the following is nearest to $\sqrt{4200}$: 20, 60, 200 or 600?

6. Which of the following is nearest to $\sqrt{0.038}$: 0.02, 0.06, 0.2, or 0.6?

7. Which is nearest to $\sqrt{0.017}$: 0.004, 0.01, 0.04 or 0.1?

Use tables or a calculator to find the square roots of the following numbers, correct to 4 sig. fig.:

8. 3, 18, 40 and 90

9. 5.6, 43, 8.25 and 55.7

10. 7.624, 76.24, 1.508 and 15.08

11. 6.52, 65.2, 652 and 6520

12. 43.6, 4.36, 0.436 and 0.0436

13. 1.82, 18.2, 182 and 1820

14. 0.009 17, 0.0917, 0.000 917 and 0.917

15. Calculate 2^3, 3^3, 4^3 and 5^3.
State the nearest integer (2, 3, 4 or 5) to $\sqrt[3]{10}$, $\sqrt[3]{70}$, $\sqrt[3]{119}$, $\sqrt[3]{25}$

16. $\sqrt[3]{90}$ is between 4 and 5. Make a similar statement for $\sqrt[3]{42}$ and for $\sqrt[3]{15}$.

17. Which of the following is nearest to $\sqrt[3]{1250}$: 5, 10, 50 or 100?

18. Which of the following is nearest to $\sqrt[3]{0.027\,89}$: 0.0003, 0.003, 0.03, 0.3?

Calculations with aids

Exercise 13

Use tables of logarithms or a calculator to find the values of the following correct to 3 sig. fig.:

1. 7.8×6.9
2. 32×9.7
3. 4.59×20.4
4. 28.3×7.92
5. 837×1.46
6. 2.08×13.6
7. $9.8 \div 2.6$
8. $42.6 \div 5.72$
9. $57.8 \div 4.7$
10. $2450 \div 33.2$
11. $684 \div 8.5$
12. $936 \div 167$
13. 3.72^3
14. 22.6^3
15. 1.65^4
16. $\sqrt[3]{9.23}$
17. $\sqrt[3]{754}$
18. $\sqrt[4]{95.8}$
19. 5.4×0.87
20. 26×0.94
21. 38×0.056
22. 83.7×0.632
23. 0.948×0.873
24. 0.047×326
25. $16.4 \div 28.9$
26. $5.7 \div 96.4$
27. $43 \div 82$
28. $0.275 \div 0.162$
29. $0.487 \div 0.89$
30. $0.652 \div 0.017$
31. 0.86^3
32. 0.92^4
33. $\sqrt[3]{0.642}$
34. $\sqrt[3]{0.083}$
35. $\sqrt[4]{0.902}$
36. $\sqrt[4]{0.076}$
37. $\dfrac{2.37 \times 5.68}{4.53}$
38. $\dfrac{54.6 \times 7.94}{86.3}$
39. $\dfrac{24.4 \times 0.618}{5.23}$
40. $\dfrac{17.3 \times 0.92}{4.63}$
41. $\dfrac{0.317 \times 18.6}{7.83 \times 0.65}$
42. $\dfrac{342}{54 \times 63}$
43. $4.2^2 \times 5.9$
44. 6.8×9.4^2
45. $\dfrac{3.7^2}{1.8}$
46. $\dfrac{8.3}{2.2^2}$
47. $\sqrt{\dfrac{16.3}{3.4}}$
48. $\sqrt[3]{\dfrac{38.5}{9.7}}$
49. $\dfrac{8.7 + 3 \times 1.54}{193 \times 6.53}$
50. $\dfrac{9.6 \times 4 - 21.8}{36.4 \times 0.67}$
51. If $x = 7.3$ and $y = 4.9$, calculate (a) xy (b) x^2 (c) y^3
52. If $x = 5.4$ and $y = 6.6$, calculate (a) xy (b) y^2 (c) x^3
53. If $c = 24.5$ and $d = 7.8$, calculate $c \div d$
54. If $r = 4.6$ and $\pi = 3.14$, calculate (a) r^2 (b) πr^2 (c) $2\pi r$

Ratio and scales

$$560\text{ g}: 2\text{ kg} = \frac{560\text{ g}}{2\text{ kg}} = \frac{560\text{ g}}{2000\text{ g}} = \frac{56}{200} = \frac{7}{25} = 7:25$$

To divide £48 among Ann, Betty and Carole in the ratio 2: 3: 7, first divide it into 12 parts (2 + 3 + 7) and then give Ann 2 parts, Betty 3 parts and Carole 7 parts. $\frac{1}{12}$ of £48 = £4 and so Ann has 2 × £4 = £8, Betty has 3 × £4 = £12 and Carole has 7 × £4 = £28.

A plan of a room has a scale of 1: 50. This means that each length on the plan is $\frac{1}{50}$ of the corresponding length in the room. 7 m in the room is represented on the plan by $\frac{1}{50}$ of 7 m = $\frac{1}{50}$ of 700 cm = 14 cm. As 1 metre is represented by 2 cm, the scale can also be stated as 2 cm to 1 m.

Exercise 14

Express each of the following ratios in its simplest form:

1. 12:18 **2.** 35:45 **3.** 8:32 **4.** 42:7
5. 10 cm:15 cm **6.** 450 m:630 m **7.** £5:£1.40 **8.** 60p:£3
9. 30 m:1 km **10.** 700 g:2 kg **11.** 3 km:420 m
12. 3 kg:360 g

13. £15 is divided between Jack and George in the ratio 2:3. What fraction does each receive? How much does each receive?

Divide the given number or quantity in the given ratio:

14. 35, 2:3 **15.** 55, 3:2 **16.** £40, 7:3 **17.** £32, 3:5
18. 8 cm, 1:3 **19.** 30 g, 4:1 **20.** 1 m, 3:7 **21.** 1 kg, 13:7

22. Divide £3 in the ratio 1:4:5.

23. A certain pastry mixture consists of flour, margarine, and sugar in the ratio 6:3:1. How much of each is needed to make 600 g of the mixture?

24. Alan, Bill and Chris shared a sum of money in the ratio 5:7:8. What fraction did each receive? If Alan received £7.50, how much did Bill receive?

25. A sum of money is divided in the ratio 3:4:5. The smallest part is £120. What is the largest part?

26. An alloy consists of metals **A** and **B** in the ratio 5:6. How much of **A** should be mixed with 30 kg of **B**?

27. A cement mix is formed by mixing cement and sand in the ratio 1:5. How much sand should be mixed with 0.3 m^3 of cement and what volume of mix is obtained?

28. A certain toothpaste is sold in tubes of three sizes–small, medium and large. The amounts in the tubes are in the ratio 1:3:5. The medium size tube contains 84 ml. What does the large size contain?

29. Find the value of x in each of the following:
(a) $x:9 = 2:3$ (b) $15:x = 3:4$
(c) $x:7 = 4:5$ (d) $3:x = 5:4$

30. Express the ratios without fractions and in their simplest forms:
(a) $1\frac{1}{2}:2\frac{1}{2}$ (b) $\frac{2}{3}:\frac{1}{2}$ (c) 0.72:0.9

31. Express each of the following in its simplest form:
(a) 0.72 kg:0.96 kg (b) 0.036 mm:0.018 mm
(c) 1.2 g:0.75 g

32. Three partners agreed to share the profits of their business as follows: **A** had a fixed salary of £5000 p.a. and **B** had £4000 p.a. The remainder was to be shared between **A, B** and **C** in the ratio 2:3:5. One year the profits were £32 000.
(a) How much remained after the two salaries had been paid?
(b) How much did each partner receive from this remainder?

33. The plan of a house has a scale of 1:100. What length in centimetres on the plan represents (a) 6 m (b) 2.4 m (c) 80 m in the house? What length in metres in the house is represented by (a) 3 cm (b) 4.2 cm (c) 6 mm?

34. The scale of a map is 5 cm to 1 km. What distance in metres is represented by (a) 1 cm (b) 3.7 cm?
What length on the map represents (c) 3 km (d) 0.7 km?

35. The scale of a map of the British Isles is 1 cm to 50 km.
(a) On the map Dublin is 9 cm from London. What is the real distance?
(b) On the map Glasgow is 3.5 cm from Belfast. What is the real distance?
(c) If Liverpool is 200 km from Newcastle, what is the distance between them on the map?
(d) If Aberdeen is 620 km from Cardiff, what is the distance between them on the map?

36. A room is 6 m by 4.5 m. A plan is drawn using a scale of 4 cm to 1 m. What lengths are used for the sides of the room on the plan?
On the plan a window is 12 cm wide and a door is 3.2 cm wide. What are the actual widths?

37. The scale of a certain Ordnance Survey map is 1:50 000.
(a) What distance in kilometres is represented by 1 cm?
(b) The distance between a station and a school is 3 cm on the map. What is the actual distance?
(c) The actual distance between Weston and Easton is 7 km. What is the distance on the map?

Proportion

Exercise 15

1. If 5 oranges cost 45p, find the cost of one orange and the cost of 8 oranges.
2. 3 copies of a book cost £13.80. Find the cost of one copy and the cost of 7 copies.
3. 3 metres of material cost £8.40. How much will 5.5 metres cost?
4. The postage for 6 parcels is £2.10. Find the cost of the postage for 8 such parcels.
5. An aircraft travels 72 km in 8 minutes. How far does it go in 1 minute? How far in 3 minutes?
6. A car travels 180 metres in 10 seconds. At the same rate, how far does it go in 7 seconds?
7. A man earns £285 in 3 weeks. How much does he earn in 8 weeks?
8. A car can travel 92 km on 12 litres of petrol. How far is it likely to go on 30 litres?
9. When 5 children share a packet of sweets each gets 8 sweets. How many are there in the packet? How many does each get when 4 children share a packet containing the same number?
10. A farm has sufficient grain to feed 40 hens for 6 days. How long would the grain last 10 hens? How long would it last 30 hens?
11. Mark takes 64 paces when walking the length of a corridor. His pace is 75 cm. How long is the corridor? Jane takes 80 paces for the same corridor. How long is her pace?
12. A bookshelf can hold 15 books of width 4 cm. How many books of width 3 cm can this shelf hold?
13. A bar of metal is 12 cm long and has a mass of 180 grams. Another bar of the same metal and same cross-section is 16 cm long. What is its mass?
14. A market gardener can grow 6000 plants on 2.4 hectares. How many can he grow on 4.2 hectares?
15. For a certain length of wire fencing a farmer decides to place 21 posts 3 metres apart. How long is the fencing? (Notice that there are 20 gaps between the posts.) How many posts would be needed if he placed the posts 4 metres apart?
16. A debt can be repaid in 30 weeks by paying £1.60 per week. How much per week must be paid to repay the debt in 20 weeks?
17. A certain quantity of liquid will fill 2000 bottles each holding 180 ml. How many bottles holding 150 ml can be filled with the same quantity of liquid?
18. 9 looms weave 675 metres of cloth in 5 hours. How much cloth does each loom weave each hour? How much will 5 looms weave in 7 hours?

19. The cost of a telephone call lasting 6 min 40 s was £1.20. What is the cost of a call lasting 8 min 20 s at the same rate?
20. A machine produces 180 articles in one hour.
(a) How many articles are produced by 5 such machines in 3 hours?
(b) How many of these machines are needed to produce 360 articles in $\frac{1}{2}$ hour?

Time and speeds

2.35 a.m. is equivalent to 02 35 hours on the 24-hour notation.
2.35 p.m. is equivalent to 14 35 hours on the 24-hour notation.

Exercise 16

1. Convert to the 24-hour notation:
6 a.m., 6 p.m., 5.40 a.m., 5.40 p.m., 9.23 a.m., 9.23 p.m.
2. Convert to the a.m./p.m. notation:
06 00 h, 16 00 h, 04 10 h, 14 10 h, 11 44 h, 21 44 h.
3. Express in both notations: (a) twelve minutes past eight in the morning (b) a quarter past four in the afternoon (c) a quarter to seven in the evening
4. A television film started at 7.40 p.m. and ended at 9.35 p.m. State its running time in minutes.
5. (a) A bus left Camford at 08 50 h and arrived at Granton at 11 12 h. How many hours and minutes did the journey take?
(b) A train left Camford at 23 40 h and arrived at Granton at 01 15 h. How many minutes did it take?
6. State the number of hours and minutes between: (a) 8.20 a.m. and 10.40 a.m., (b) 11.45 a.m. and 1.00 p.m., (c) 7.50 p.m. and midnight, (d) 10.55 p.m. and 2.05 a.m.
7. State the number of hours and minutes between: (a) 05 30 h and 09 40 h, (b) 10 15 h and 12 05 h, (c) 14 50 h and 20 10 h, (d) 18 40 h and 02 30 h.
8. A man started a night shift at 20 30 h and worked for 7 h 45 min. At what time did he finish?
9. There was a high tide at Brightsea one Monday at 09 17 h. The time between consecutive high tides is 12 h 25 min. When were the next two high tides?

10. Which of the following were leap years: 1944, 1953, 1956, 1966, 1970, 1972?
11. Find the number of days between the two given dates in each of the cases below. Do NOT include the first and last days. For example, between 8 August and 11 August there are two days.
(a) 5 September and 10 September (b) 28 January and 5 February
(c) 7 March and 20 April (d) 30 March and 2 May
January and March have 31 days each and April has 30.
12. A record takes 2 minutes 40 seconds to play at a speed of 45 revolutions per minute. How many revolutions does it make?
13. The first part of a concert will last for 1 h 5 min. There is to be an interval of 15 minutes and the second part will last for 35 min. The concert will start at 7.30 p.m. When will it end?
14. A man boarded a ship at 10.45 a.m. on a Friday and left it 75 h 35 min later. At what time and on what day did he leave the ship?

Speed

$$\textbf{Average Speed} = \frac{\textbf{Distance travelled}}{\textbf{Time taken}}$$

Suppose that an aircraft travels 300 km in 25 min. 25 min = $\frac{5}{12}$ hour.

The average speed = $\frac{300}{\frac{5}{12}}$ km/h = 300 × $\frac{12}{5}$ km/h = 720 km/h.

In 1 hour it travels 720 km and so in 4 hours it travels
4 × 720 km = 2880 km. This illustrates that

$$\textbf{Distance} = \textbf{Speed} \times \textbf{Time}$$

The time taken to travel 720 km is 1 hour,
the time taken to travel 1 km is $\frac{1}{720}$ hour and so
the time taken to travel 540 km is $\frac{540}{720}$ h = $\frac{3}{4}$ h = 45 min.
This illustrates that

$$\textbf{Time taken} = \frac{\textbf{Distance}}{\textbf{Speed}}$$

Exercise 17

1. An aircraft flies at 720 km/h. How far does it go in (a) 3 h (b) 1 min (c) 8 min?
2. A train is travelling at 90 km/h. How far does it go in
(a) $\frac{1}{2}$ h (b) 1 min (c) 12 min?
3. Find the speed in each of the following cases:
(a) An aircraft travels 1940 km in 2 h. (km/h)
(b) A boy runs 60 m in 8 s. (m/s)
(c) A boat travels 7 kilometres in 20 min. (km/h)

4. Find the speed in each of the following cases: (a) A train travels 9 km in 4 min. (km/h) (b) A man swims 11 m in 5 s. (m/s) (c) A spacecraft travels 32 km in 3 s. (km/h)
5. At 18 km/h how long does it take a boat to travel (a) 36 km (b) 6 km (c) 12 km?
6. At 120 km/h how long does it take a train to travel (a) 60 km (b) 2 km (c) 14 km?
7. A car is travelling at 72 km/h. How many metres does it travel in (a) 1 minute (b) 1 second? What is its speed in m/s?
8. A boy runs 8 m in 1 s. At the same rate (a) how far would he go in metres in 1 minute and (b) how far would he go in kilometres in 1 hour? State his speed in km/h.
9. An athlete ran 400 m in 45 s. Find his average speed in km/h.
10. A train travelled 120 km at an average speed of 45 km/h. How long did the journey take?
11. To reach a farm I travelled 20 km by bus at an average speed of 30 km/h and then walked $1\frac{1}{2}$ km at an average speed of 6 km/h. How long did the journey take?
12. A journey took 1 h 40 min at an average speed of 96 km/h. How long was the journey?
13. A train left Edinburgh at 09 04 and arrived at Glasgow at 10 34. Find (a) the time for the journey and (b) the average speed if the distance was 75 km.
14. A train travelled for 20 km at 120 km/h and then for 25 km at 100 km/h. Find (a) the total distance travelled (b) the total time taken and (c) the average speed.
15. A girl cycled 12 km in 44 minutes. Calculate her average speed in m/s, correct to 1 dec. pl.
16. An aircraft travelled 240 km in 16 min. Find its average speed in km/h.
17. An aircraft flies at 1200 kilometres per hour.
    (a) At this speed, how far does it travel between 13 50 and 15 10?
    (b) How many seconds does it take to travel 1 kilometre?
18. A train left Ayton at 23 15 and arrived at Beeford at 00 28. How long did it take? The distance between the towns is 136 km. Calculate the average speed, correct to the nearest km/h.
19. A delivery van left its depot at 09 15 and travelled to a factory 105 km away at an average speed of 63 km/h. At what time did it arrive?
    It started its return journey at 12 45 and arrived back at its depot at 14 15. How long did the return journey take and what was the average speed?

Rectangles and triangles

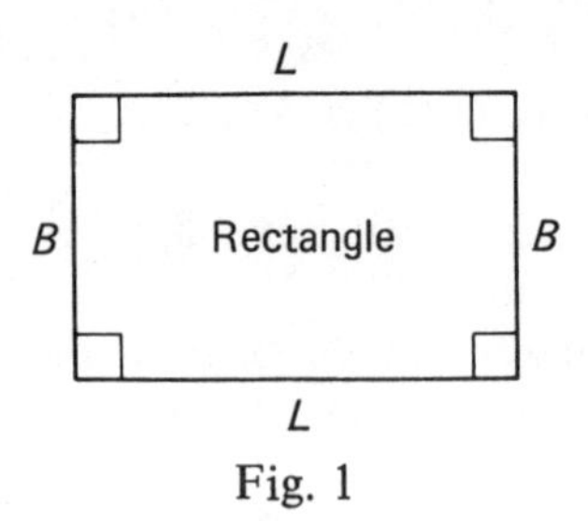

Fig. 1

For a rectangle

Area = Length × Breadth $A = L \times B$

Length = $\dfrac{\text{Area}}{\text{Breadth}}$ $L = \dfrac{A}{B}$

Breadth = $\dfrac{\text{Area}}{\text{Length}}$ $B = \dfrac{A}{L}$

Perimeter (distance round it) = 2 × Length + 2 × Breadth $P = 2L + 2B$

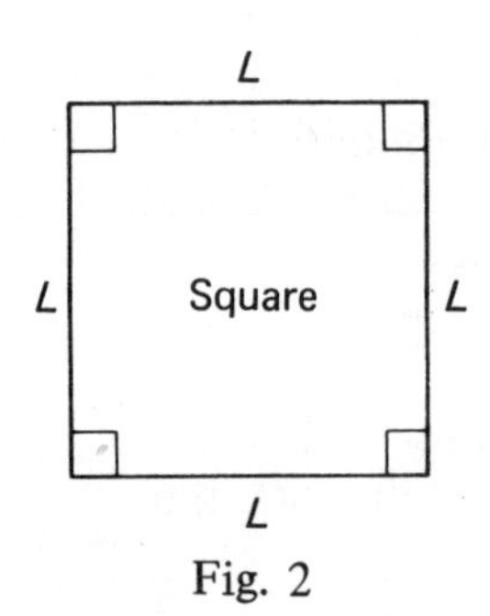

Fig. 2

For a square

Area = $(\text{Length})^2$ $A = L^2$

Length = $\sqrt{(\text{Area})}$ $L = \sqrt{A}$

Perimeter = 4 × Length $P = 4L$

Exercise 18

Calculate the perimeter and area of each of the following rectangles:

1. Length 6 cm, breadth 4 cm **2.** Length 10 cm, breadth 9 cm

3. Length 8 m, breadth 3.5 m **4.** Length 9.5 m, breadth 6 m

5. Length 3.2 m, breadth 2.4 m **6.** Length 7.6 m, breadth 5.3 m

Calculate the perimeter and area of each of the following squares:

7. Side 7 cm **8.** Side 10 cm **9.** Side 3.5 m **10.** Side 4.2 m

11. The perimeter of a square lawn is 36 m. Calculate the length and area.

12. A square has an area of 36 m^2. Calculate its length and perimeter.

13. A rectangle has a length of 12 cm and an area of 84 cm^2. State its breadth.

14. A rectangle has a breadth of 4 cm and an area of 26 cm^2. Find its length and its perimeter.

15. The area of a rectangle is 48 cm^2 and the breadth is 6.4 cm. Calculate the length.

16. The area of a rectangle is 119 cm^2 and the breadth is 8.5 cm. Calculate the breadth.

17. A rectangular piece of card, 20 cm by 16 cm, is cut into squares of side 4 cm. How many squares are obtained?
18. A room is 5 m by 4 m. How many square tiles of side $\frac{1}{2}$ m are needed to cover the floor?
19. Express in square millimetres: (a) 1 cm^2 (b) 4 cm^2 (c) 0.4 cm^2 (d) 4.6 cm^2
20. Express in square centimetres: (a) 1 m^2 (b) 6 m^2 (c) 0.6 m^2 (d) 4.8 m^2
21. Express in square metres: (a) 3 are (b) 5.3 a (c) 1 ha (d) 5 ha
22. (a) Express in square metres (i) 64 000 cm^2 (ii) 758 cm^2
(b) Express in hectares (i) 30 000 m^2 (ii) 78 000 m^2
23. A rectangle is 85 mm by 24 mm. Calculate its area in cm^2.
24. A table cloth is 160 cm by 125 cm. Calculate its area in m^2.
25. A rectangular field is 360 m by 200 m. Calculate its perimeter in kilometres and its area in hectares.
26. A rectangular field is 350 m by 400 m. Calculate its perimeter in kilometres and its area in hectares.
27. Calculate the shaded area in Fig. 3. (It is half the area of the rectangle.)

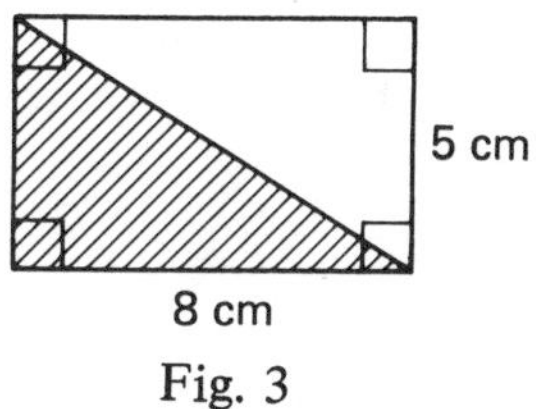

Fig. 3

28. In Fig. 4, calculate the area A, the area B and the area of the shaded triangle.

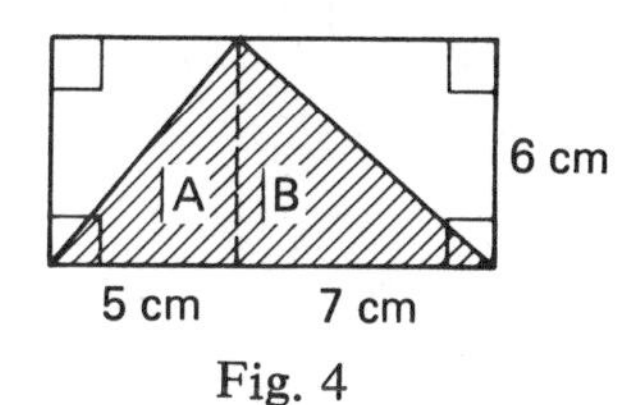

Fig. 4

29. Two squares have sides of 4 cm and 6 cm. State the ratio of (a) the sides (b) the perimeters (c) the areas.
30. A carpet costing £2.20 per m^2 is laid in a room 6 m by 5 m. Find the area of the carpet in m^2 and the cost.
31. A picture is mounted on a peice of card to leave a border as shown in Fig. 5. Calculate the area of
(a) the picture and border together (large rectangle),
(b) the picture (small rectangle),
(c) the border.

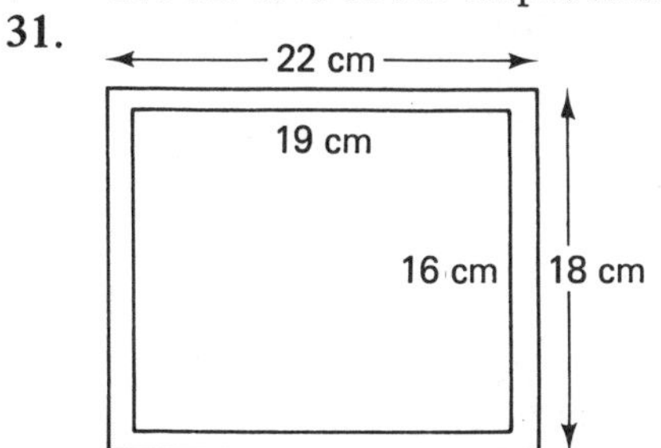

Fig. 5

32. 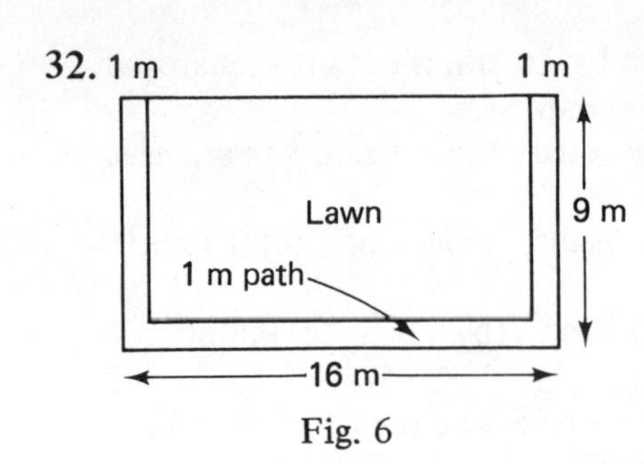

Fig. 6

Fig. 6 shows a lawn with a path on three sides. Calculate (a) the length and breadth of the lawn (b) the area of the lawn (c) the area of the path.

33. A room is 8.5 m long, 6.5 m wide and 3.4 m high. The walls are to be papered using rolls of paper of length 11 m and width 50 cm.
(a) Find the perimeter of the room. (b) Find the area of the walls, ignoring doors and windows. (c) Find the area of one roll of paper.
(d) Estimate the number of rolls needed.

Triangles and trapezia
Area of a traingle = $\frac{1}{2}$ (base × height)

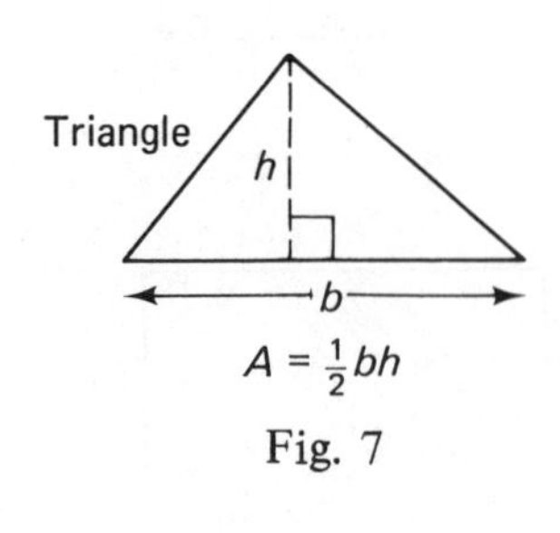

Fig. 7

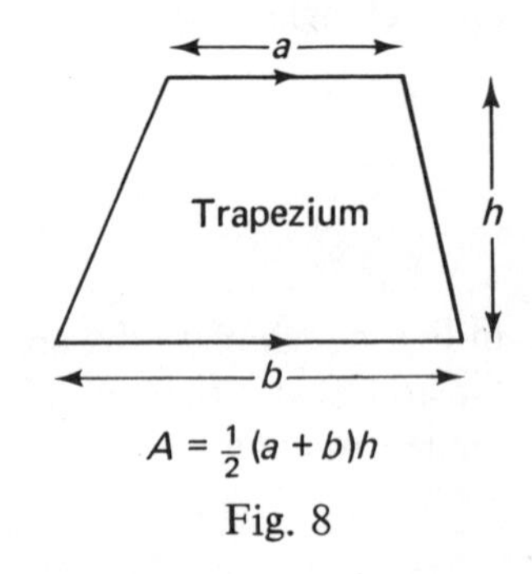

Fig. 8

Area of a trapezium = $\frac{1}{2}$ (sum of the parallel sides) × distance between them.

Exercise 19

In questions **1** to **6** calculate the area of each triangle, the measurements being in centimetres.

1. Fig. 9

2. 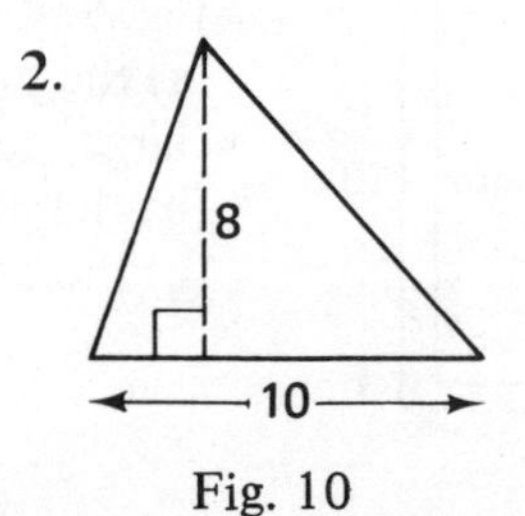

Fig. 10

3. 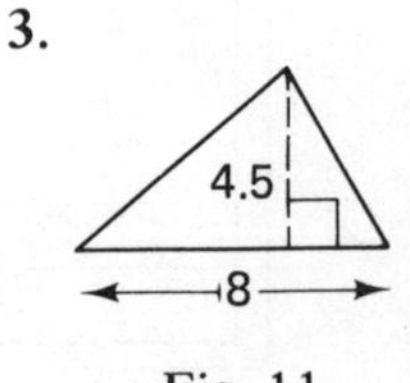

Fig. 11

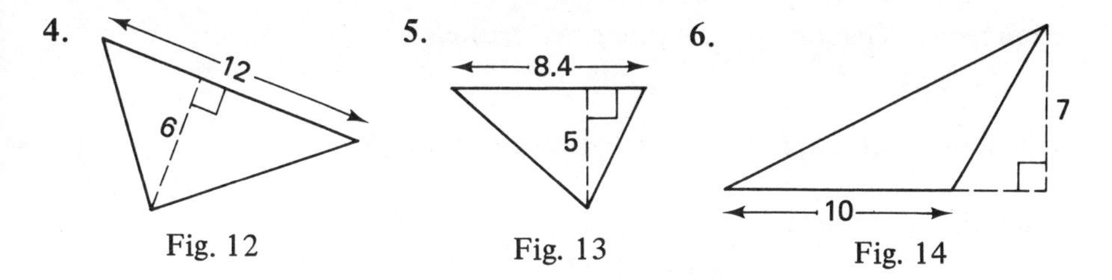

Fig. 12 Fig. 13 Fig. 14

7. The area of a triangle is 27 cm^2. Its base is 9 cm. Find its height.
8. The area of a triangle is 70 cm^2. Its height is 10 cm. Find its base.
9. The area of a traingle is 57 cm^2. Its height is 12 cm. Find its base.
10. The area of a traingle is 18 cm^2. Its base is 7.5 cm. Find its height.

In questions **11**, **12** and **13** find the area of each figure by dividing it into rectangles and triangles. The measurements are in centimetres.

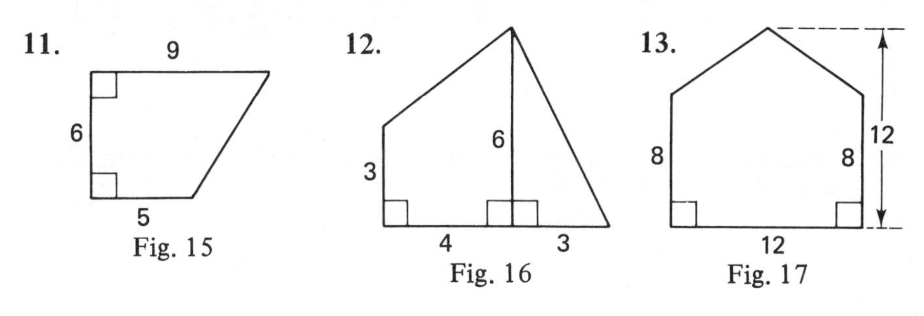

Fig. 15 Fig. 16 Fig. 17

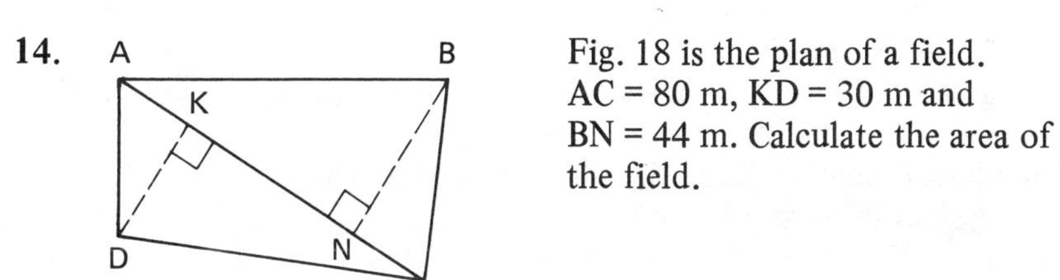

Fig. 18

Fig. 18 is the plan of a field. AC = 80 m, KD = 30 m and BN = 44 m. Calculate the area of the field.

In questions **15**, **16** and **17** find the area of each trapezium. The measurements are in centimetres.

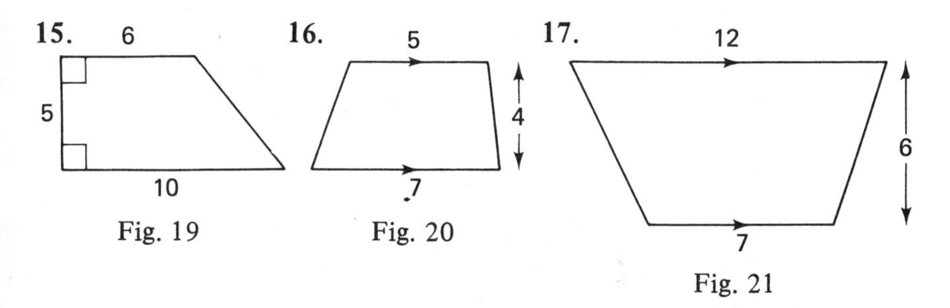

Fig. 19 Fig. 20 Fig. 21

Pythagoras' Theorem for a right-angled triangle

$a^2 = b^2 + c^2$

(the longest side, a, is called the hypotenuse.)

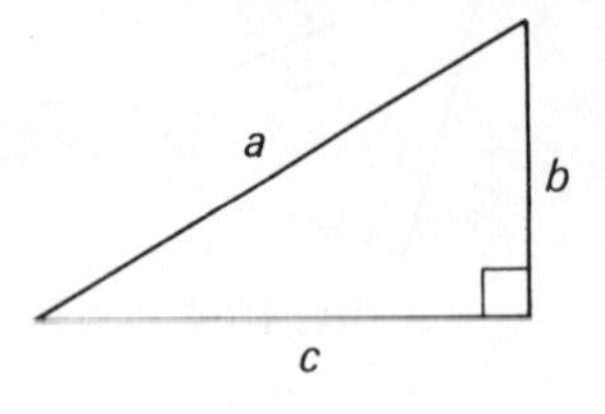

Fig. 22

Example: If $b = 6$ and $c = 7$, find a.

$a^2 = 6^2 + 7^2 = 36 + 49 = 85$

$a = \sqrt{85} \simeq 9.2$

Exercise 20

In questions **1** to **4** you are given two sides of a right-angled triangle. Calculate the hypotenuse:

1. 8 cm and 6 cm **2.** 12 cm and 5 cm **3.** 8 cm and 15 cm

4. 18 cm and 24 cm

5. A rectangle has sides of 16 cm and 12 cm. Calcualte the length of a diagonal.

6. A football pitch has sides of 120 m and 90 m. What is the length of a diagonal?

In questions 7 to **10** you are given the two sides next to the right-angle in a right-angled triangle. Calculate the length of the hypotenuse, correct to 1 dec. pl.:

7. 7 cm and 5 cm **8.** 11 cm and 9 cm **9.** 6.4 cm and 7.3 cm

10. 9.6 cm and 8.3 cm

11. A ship sailed 10 nautical miles South and then 7 nautical miles East. How far was it then from its starting point?

12. Calculate, correct to 2 sig. fig., the length of a diagonal of a rectangle having sides of 8 cm and 7 cm.

13. Calculate the length of the diagonal of a square of side 10 cm.

14. A flag pole 12 m high is to be held by wire stays attached to it 4.5 m from its top and to the ground 4 m from the base. (Fig. 23) Find the length of each stay.

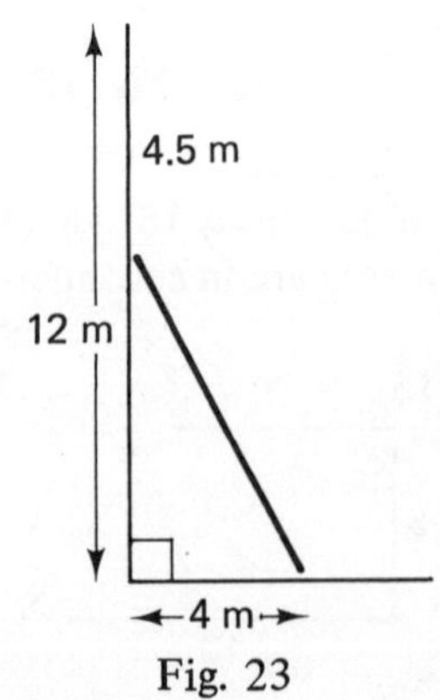

Fig. 23

In questions **15** to **18** you are given the hypotenuse and another side of a right-angled triangle. Calculate the third side.

15. 10 cm and 6 cm
16. 26 cm and 10 cm
17. 9 cm and 5 cm (to 2 sig. fig.)
18. 11 cm and 7 cm (to 2 sig. fig.)
19. In triangle ABC, angle A is 90°, AB = AC and BC = 10 cm. Calculate AB, correct to 2 sig. fig.
20. The diagonal of a square is 8 cm. Calculate the length of a side.

Circles

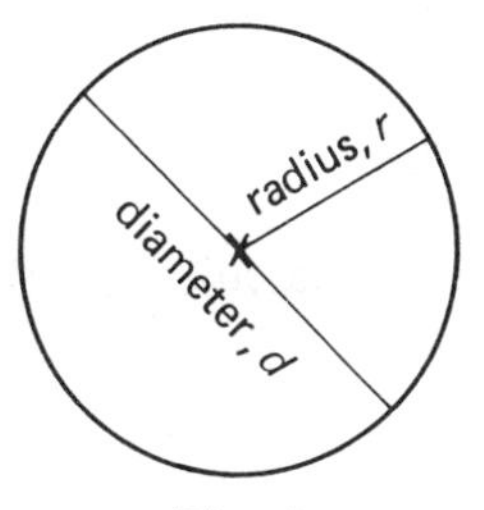

Fig. 1

The circumference, c, is given by

$$c = \pi d$$

or $$c = 2\pi r$$

The area, A, is given by $A = \pi r^2$
(This means $\pi \times r \times r$)
π is approximately 3.14 or $\frac{22}{7}$

Example 1: Calculate the circumference of a circle of radius 14 cm

$r = 14, \quad d = 28$
$c = \pi d \simeq \frac{22}{7} \times \frac{28}{1} = 88$

The circumference is 88 cm.

Example 2: Calculate the area of a circle of radius 6 cm.

$A = \pi r^2 \simeq 3.14 \times 6 \times 6 = 113.04$

The area is 110 cm^2, to 2 sig. fig.

Exercise 21

Using $\frac{22}{7}$ for π, calculate the circumference of each circle:

1. Diameter 7 cm
2. Radius 21 cm
3. Radius 2.8 cm

Using 3.14 for π, calculate the circumference of each circle, to 2 sig. fig.:

4. Diameter 3 cm
5. Diameter 4 cm
6. Diameter 15 cm
7. Radius 4 cm
8. Radius 10 cm
9. Radius 6.8 cm

Using $\frac{22}{7}$ for π, calculate the area of each circle:

10. Radius 7 cm
11. Radius 14 cm
12. Diameter 7 cm

Using 3.14 for π, calculate the area of each circle, to 2 sig. fig.:

13. Radius 4 cm **14.** Radius 10 cm **15.** Radius 8.4 cm

16. The diameter of a car wheel is 75 cm. How many metres, to 2 sig. fig., does the car go when the wheel makes 100 revolutions?

17. Some thin wire is wound onto a cylindrical drum of diameter 65 mm. Calculate, to 2 sig. fig., the length in cm of 1 turn and of 200 turns.

18. A running track has two semi-circular ends of radius 28 m and two straight sides. The perimeter of the track is 400 m. Using $\frac{22}{7}$ for π, calculate (a) the perimeter of each curve (b) the length of each straight.

19. Water from a rotating sprayer is thrown over a circular area of radius 6 m. Calculate the area covered, correct to the nearest square metre.

20. There is a path of width 1 m round a circular pond of radius 4 m. Calculate the area of (a) the pond (b) the pond and path together (c) the path.

21. Calculate the diameter of a circle of circumference 66 cm. ($\pi \triangleq \frac{22}{7}$, $d = c/\pi$)

22. The circumference of a circle is 216 m. Calculate its diameter, correct to 2 sig. fig., using $\pi \triangleq 3.14$.

23. The area of a circle is 154 cm^2. Using $\frac{22}{7}$ for π, calculate $(\text{radius})^2$ and hence the radius. ($r^2 = \dfrac{A}{\pi}$)

24. Calculate, correct to 2 sig. fig., the radius of a circle of area 910 cm^2, using $\pi \triangleq 3.14$.

25. A circle is cut from a square sheet of metal as shown in Fig. 2. State the radius of the circle.
Calculate the area of (a) the square sheet of metal (b) the circle (c) the metal wasted (the shaded part.)

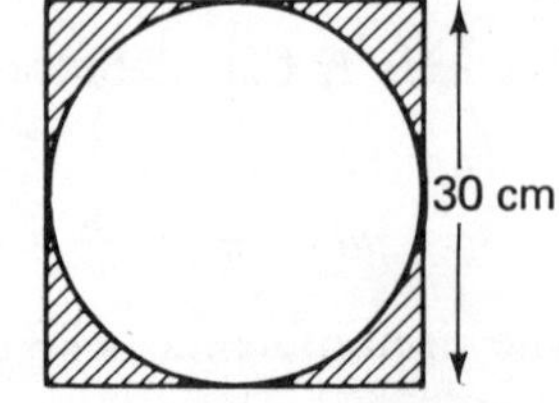

Fig. 2

26. The circumference of a circle is 54 metres. Calculate (a) the diameter (b) the area.
Use 3.14 for π. $d = \dfrac{c}{\pi}$

Prisms, cylinders, cones and spheres

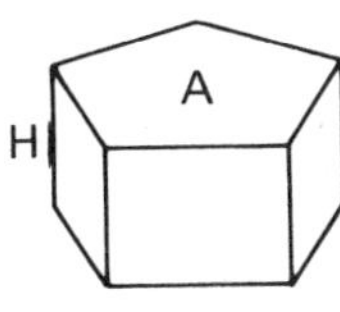

Fig. 1

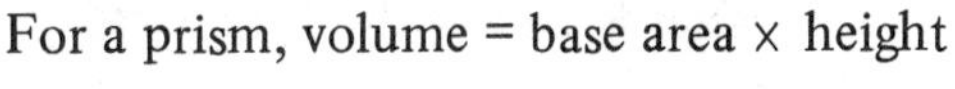

For a prism, volume = base area × height

$V = AH$

A cuboid is a prism with a rectangular base. Volume of a cuboid = length × breadth × height

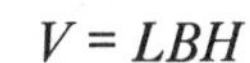

$V = LBH$

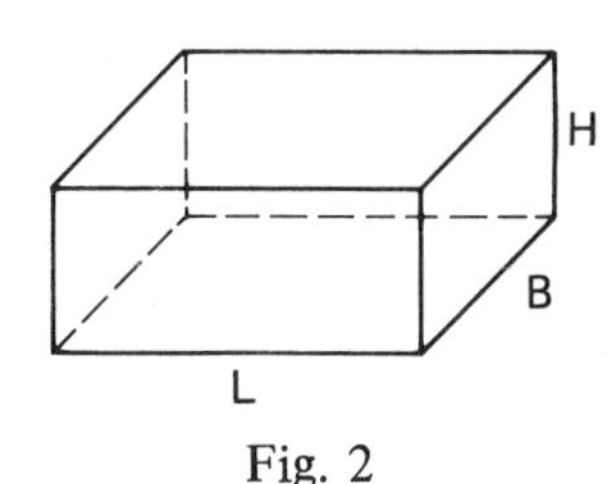

Fig. 2

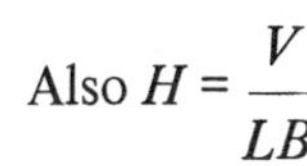

Also $H = \dfrac{V}{LB}$

Exercise 22

Calculate the volume of each of the following solids:

1. A cuboid of length 4 cm, breadth 3 cm and height 2 cm.
2. A cuboid of length 10 cm, breadth 6 cm and height 4 cm.
3. A cube of edge 3 cm.
4. A cube of edge 5 cm.
5. A prism having a base area of 9 cm^2 and a height of 4 cm.
6. A prism having a base area of 13 cm^2 and a height of 6 cm.
7. A cuboid having a square base of side 7 cm and a height of 10 cm.
8. A cuboid having a square base of side 10 cm and a height of 7 cm.
9. A rectangular box is 20 cm by 15 cm by 6 cm. Find (a) the area of the largest face (b) the total surface area.
10. Find the total surface area of a cuboid of length 8 cm, breadth 6 cm and height 5 cm.
11. Find the total surface area of a cube of edge 4 cm.
12. The total surface area of a cube is 150 cm^2. Find the length of an edge.
13. How many cubes of edge 3 cm can be fitted into a box 12 cm by 9 cm by 6 cm?
14. How many cubes of edge 1 cm can be fitted into a cubical box of edge 1 m? How many cubic centimetres equal 1 cubic metre?
15. A glass tank is 30 cm long, 20 cm wide and 10 cm high. Calculate its volume in (a) cm^3 (b) millilitres (c) litres.
16. A rectangular tank is 60 cm by 50 cm by 40 cm. How many litres of water will it hold?
17. 240 cubes of edge 1 cm can be packed into a cardboard box. The base of the box is 8 cm by 6 cm. How many cubes are there in each layer? How many layers? What is the height of the box?

18. A cuboid has a volume of 252 cm^3, a length of 9 cm and a width of 7 cm. Calculate its height.

19. 12 000 cm^3 of water are poured into an empty fish tank of length 40 cm and width 25 cm. What is the area of the base of the tank? What is the depth of the water?

20. Calculate the volume of a cuboid having (a) length 6.5 cm, breadth 4 cm and height 2.5 cm. (b) length 9.5 cm, breadth 6.4 cm and height 5 cm.

21. Calculate the volume of a cuboid having a square base of side 5.5 cm and a height of 8 cm.

22. How many cubes of edge 3 cm can be made from 1000 cm^3 of metal? (There will be some metal left over.)

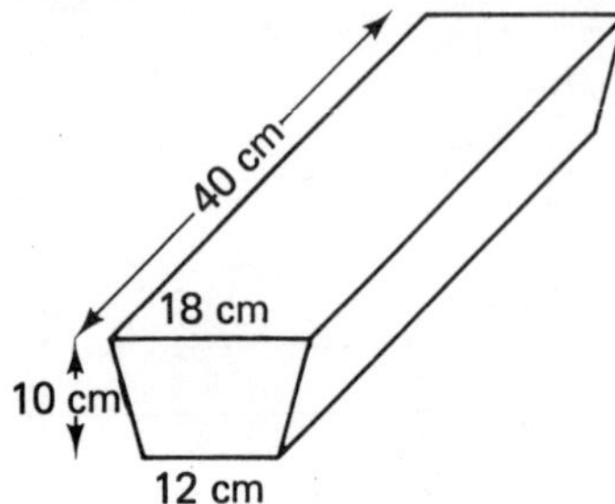

Fig. 3

Fig. 3 shows a trough with a rectangular base, each end being a trapezium. Calculate (a) the area of an end (b) the volume of the trough.

24.

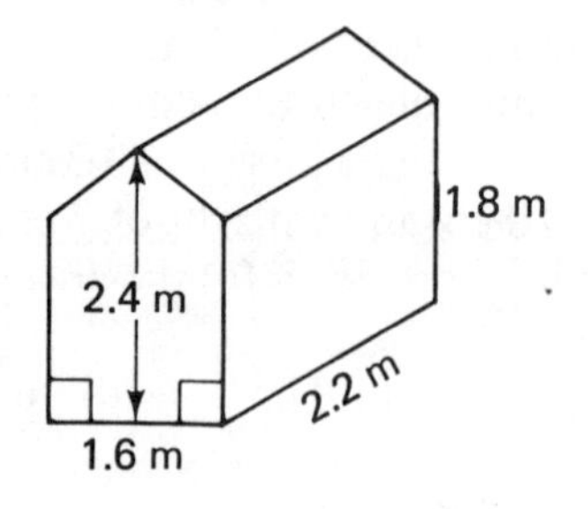

Fig. 4

Fig. 4 shows a garden shed. Calculate, to 2 sig. fig., (a) the area of an end (b) the volume.

25. A tank is 60 cm by 50 cm. Water runs in at the rate of 1 litre per minute. How long does it take for the level to rise by 4 cm? (1 litre = 1000 cm^3.)

26. Fig. 5 shows a rectangular swimming bath which is 50 m long and 30 m wide. The depth of the water increases uniformly from 1 m to 3 m. Calculate (a) the area of the shaded side (b) the volume of the water.

The bath is filled by a pipe of cross sectional area 0.01 m^2. If the water flows at 2 m/s, how many m^3 are discharged in 1 hour? How long does it take to fill the bath?

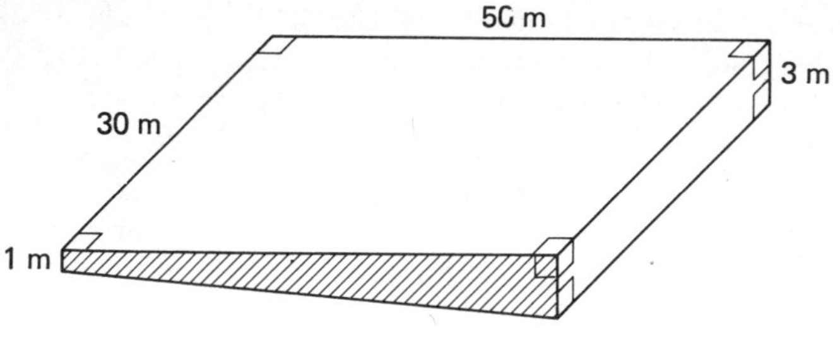

Fig. 5

27. (a) Calculate the area of a flat roof 12 m by 4 m.
(b) During a storm, 3 cm of rain fell on the roof. Calculate the volume in m^3.
(c) The rain ran into a rectangular tank of base 1.5 m by 0.8 m. Calculate the rise in the level of the water.

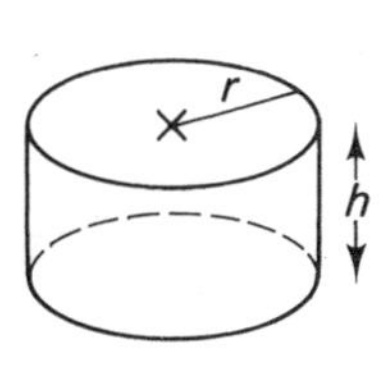

Fig. 6

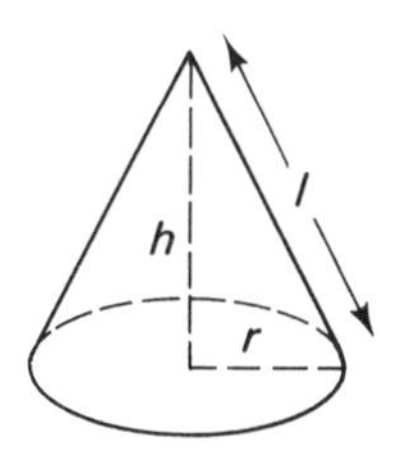

Fig. 7

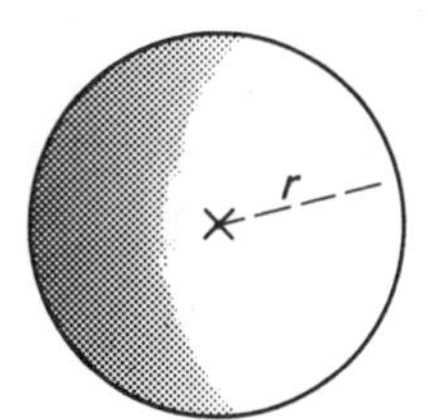

Fig. 8

Cylinder
Volume: $\pi r^2 h$
Curved surface area: $2\pi rh$

Cone
h is the vertical height
l is the slant height
Volume: $\frac{1}{3}\pi r^2 h$
Curved surface area: πrl

Sphere
Volume: $\frac{4}{3}\pi r^3$
Surface area: $4\pi r^2$

Example: Find the volume of a sphere of radius 9 cm, correct to 2 sig. fig.

$$V = \frac{4}{3}\pi r^3 = \frac{4}{3} \times 3.14 \times 9 \times 9 \times 9 \simeq 3052$$

The volume is 3100 cm^3, to 2 sig. fig.

Exercise 23

Calculate the volume of each of the following solids. When using 3.14 for π, give your answer correct to 2 sig. fig.

1. Cylinder, radius 7 cm, height 10 cm. Use $\frac{22}{7}$ for π.
2. Cylinder, radius 14 cm, height 5 cm. Use $\frac{22}{7}$ for π.
3. Cylinder, radius 5 cm, height 8 cm. Use 3.14 for π.
4. Cylinder, radius 2 cm, height 5 cm. Use 3.14 for π.

5. Cylinder, radius 4 cm, height 9 cm. Use 3.14 for π.
6. Cylinder, radius 2.5 cm, height 8 cm. Use 3.14 for π.
7. Cone, radius 3 cm, vertical height 14 cm. Use $\frac{22}{7}$ for π.
8. Cone, radius 7 cm, vertical height 10 cm. Use $\frac{22}{7}$ for π.
9. Cone, radius 4 cm, vertical height 10 cm. Use 3.14 for π.
10. Cone, radius 5 cm, vertical height 12 cm. Use 3.14 for π.
11. Sphere, radius 3 cm. Use 3.14 for π.
12. Sphere, radius 2.6 cm. Use 3.14 for π.

Calculate the curved surface area of each of the following solids. When using 3.14 for π, give your answer correct to 2 sig. fig.

13. Cylinder, radius 7 cm, height 10 cm. Use $\frac{22}{7}$ for π.
14. Cylinder, radius 2 cm, height 3.5 cm. Use $\frac{22}{7}$ for π.
15. Cylinder, radius 5 cm, height 8 cm. Use 3.14 for π.
16. Cylinder, radius 6 cm, height 8 cm. Use 3.14 for π.
17. Cone, radius 5 cm, slant height 14 cm. Use $\frac{22}{7}$ for π.
18. Cone, radius 6 cm, slant height 18 cm. Use 3.14 for π.
19. Sphere, radius 3.5 cm. Use $\frac{22}{7}$ for π.
20. Sphere, radius 4.4 cm. Use 3.14 for π.

In questions **21** to **28** use 3.14 for π and give each answer correct to 2 sig. fig.

21. A cylinder has a radius of 4.6 cm and a curved surface area of 220 cm^2. Calculate the height.

$$\left[h = \frac{A}{2\pi r}\right]$$

22. A cylinder has a height of 9.5 cm and a curved surface area of 190 cm^2. Calculate the radius.

$$\left[r = \frac{A}{2\pi h}\right]$$

23. A cylinder has a radius of 5.2 cm and a volume of 730 cm^3. Calculate the height.

$$\left[h = \frac{V}{\pi r^2}\right]$$

24. A cylinder has a height of 11.3 cm and a volume of 1590 cm^3. Calculate the radius.

$$\left[r = \sqrt{\frac{V}{\pi h}}\right]$$

25. Calculate the radius of a sphere having a surface area of 440 cm^2.

$$\left[r = \sqrt{\frac{A}{4\pi}}\right]$$

26. Calculate the radius of a sphere having a volume of 82 cm^3.

$$\left[r = \sqrt[3]{\frac{3V}{4\pi}}\right]$$

27. A cone has a radius of 4.6 cm and a volume of 200 cm^3.
Calculate its height.

$$\left[h = \frac{3V}{\pi r^2}\right]$$

28. A cone has a height of 6.2 cm and a volume of 94 cm^3.
Calculate its radius.

$$\left[r = \sqrt{\frac{3V}{\pi h}}\right]$$

29. A food can is a cylinder of diameter 3.5 cm and a height 8 cm.
Calculate the area of a label to cover the curved surface.

30. A closed cylindrical tank has a height of 1.8 m and a diameter of 0.6 m. Calculate (a) the total surface area in m^2 (b) the capacity in litres.

31.

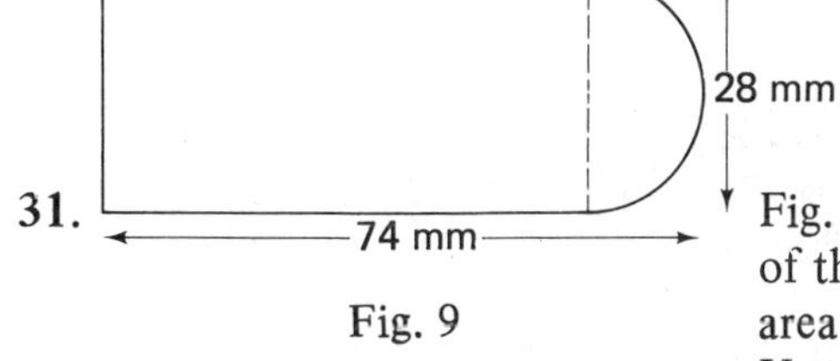

Fig. 9

Fig. 9 shows the top of a metal plate of thickness 2 mm. Calculate (a) the area of the top (b) the volume. Use $\frac{22}{7}$ for π.

32. A boiler consists of a cylinder with a hemisphere at each end. The diameter of the boiler is 140 cm and its total length is 240 cm. Calculate (a) the length of the cylindrical part (b) the volume of the cylindrical part (c) the volume of each hemisphere (d) the number of litres the boiler will hold, correct to the nearest litre. Use $\frac{22}{7}$ for π.

33. Calculate the volume of a cone of height 8.65 cm and radius 2.15 cm.

34. Calculate, in cm^3, the volume of 300 m of wire of radius 0.8 mm.

35. Calculate, correct to 3 sig. fig., the volume of a metal cone of height 18.4 cm and base radius 5.2 cm.
The cone is melted down and the metal is used to make spheres of radius 1 cm. Calculate the volume of each sphere and the number of spheres which can be made.

Approximate measurements

If the length of a line is stated to be 7 cm, correct to the nearest centimetre, the **true length** is nearer to 7 cm than to 6 cm or 8 cm. It is somewhere between 6.5 cm and 7.5 cm. 6.5 cm is the **lower limit of measurement** and 7.5 cm is the **upper limit of measurement**. The difference between the true length and the stated length is called the **error**. It can be as much as 0.5 cm. 0.5 cm is called the **absolute error**. The absolute error is half the unit of measurement, which is 1 cm in this case.

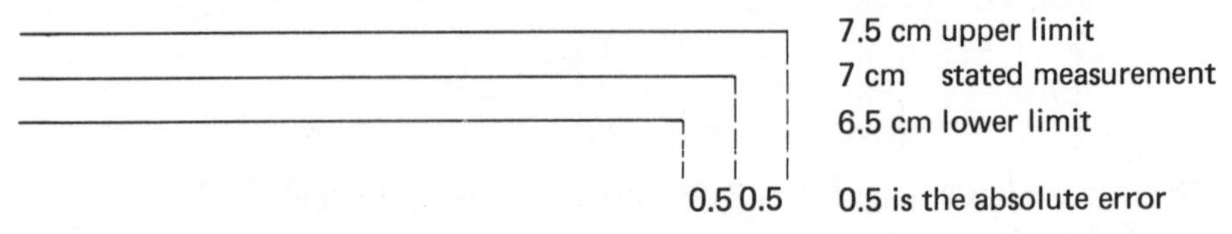

A distance is given as 9.3 km, correct to the nearest 0.1 km. The unit of measurement is 0.1 km. The absolute error is $\frac{1}{2}$ of 0.1 km = 0.05 km. The true distance is between 9.25 km and 9.35 km.

Exercise 24

1. Give the following lengths correct to the nearest cm: 6.3 cm, 5.8 cm, 23.46 cm, 38.72 cm
2. Give the following masses correct to the nearest kg: 44.6 kg, 56.3 kg, 28.23 kg, 21.74 kg
3. Each of the following masses is correct to the nearest kg. State the upper and lower limit for each: 7 kg, 5 kg, 26 kg, 30 kg
4. Each of the following lengths is correct to the nearest cm. State the upper and lower limit for each. 8 cm, 9 cm, 54 cm, 40 cm.
5. Each of the following sums of money is correct to the nearest £10: £60, £80, £240, £700. State the upper and lower limit for each.
6. Each of the following masses is correct to the nearest 10 g: 70 g, 90 g, 130 g, 600 g. State the upper and lower limit for each.
7. Copy and complete the following table:

Measurement	Unit of measurement	Absolute error	Upper limit of measurement	Lower limit of measurement
230 g	10 g	5 g	235 g	.....
47 m	1 m	0.5 m	.....	.....
6.8 cm	0.1 cm	.....	.....	.....
$8\frac{1}{2}$ hours	$\frac{1}{2}$ hour	$\frac{1}{4}$ hour	.....	.....

8. Draw up a table as in question 7 for the following measurements:

390 g to the nearest 10 g, 12 cm to the nearest 1 cm, 7.2 km to the nearest 0.1 km, 0.48 mm to the nearest 0.01 mm, $6\frac{1}{2}$ h to the nearest $\frac{1}{2}$ h.

9. State the upper and lower limits for each of the following:
(a) £5400 to the nearest £100 (b) 13.2 s to the nearest 0.1 s
(c) 9 h to the nearest $\frac{1}{2}$ h.

10. State the upper and lower limits for each of the following:
(a) 28 000 km to the nearest 1000 km (b) 46 million to the nearest million (c) 9.8 s to the nearest 0.2 s.

11. The length and breadth of a rectangle are stated to be 9 cm and 6 cm correct to the nearest cm. State the upper and lower limit for each measurement.
The upper limit for the perimeter is $9.5 + 6.5 + 9.5 + 6.5 = 32$ cm. What is the lower limit?
For the stated measurements, the area is $9 \times 6 = 54$ cm^2. The upper limit for the area is $9.5 \times 6.5 = 61.75$ cm^2. What is the lower limit for the area?

12. The length and breadth of a rectangular field are 460 m and 340 m, correct to the nearest 10 m. Calculate the upper and lower limits for the area.

13. The length of a side of a square is stated to be 14 cm to the nearest cm. Calculate the upper and lower limits for (a) the perimeter (b) the area.

14. The area of a rectangle is stated to be between 40 m^2 and 50 m^2. The length is stated to be between 8 m and 10 m. Calculate $40 \div 8$, $40 \div 10$, $50 \div 8$ and $50 \div 10$. Which is the greatest possible value, in metres, of the width? Which is the least possible value of the width?

15. The area of a rectangle is 6.4 cm^2, correct to 1 dec. pl. and the length is 2.7 cm, correct to 1 dec. pl.
(a) Which of the following gives the greatest possible value, in centimetres, of the width: $6.45 \div 2.75$, $6.35 \div 2.65$, $6.35 \div 2.75$, $6.45 \div 2.65$?
(b) Which of the above gives the least possible value of the width?

16. The area of a rectangle is 9.3 cm^2, correct to 1 dec. pl. and the width is 1.8 cm, correct to 1 dec. pl.
(a) Which of the following gives the greatest possible value of the length: $9.25 \div 1.75$, $9.25 \div 1.85$, $9.35 \div 1.75$, $9.35 \div 1.85$?
(b) Which of the above gives the least possible value of the length?

17. Robert estimates that he has cycled between 48 km and 60 km and that it has taken him between 3 and 4 hours. Calculate $48 \div 3$, $48 \div 4$, $60 \div 3$, $60 \div 4$. Which gives the greatest possible value, in km/h, for his average speed? Which gives the least?

18. The distance between two posts is 77 m to the nearest metre and Jennie runs from one to the other in 9 s to the nearest second.
(a) Which of the following gives the greatest possible value, in m/s, for her speed: $77.5 \div 9.5$, $77.5 \div 8.5$, $76.5 \div 9.5$, $76.5 \div 8.5$?
(b) Which of the above gives the least possible speed?

Percentages, discount, profit and loss

A percentage is a fraction with a denominator of 100. 7% means $\frac{7}{100}$.

$55\% = \frac{55}{100} = \frac{11}{20}$; $67\frac{1}{2}\% = \dfrac{67\frac{1}{2}}{100} = \dfrac{67\frac{1}{2} \times 2}{100 \times 2} = \dfrac{135}{200} = \dfrac{27}{40}$; $6.8\% = \dfrac{6.8}{100} = 0.068$

Notice that, to convert a percentage to a fraction or a decimal, we divide by 100. Hence we change a fraction or a decimal into a percentage by multiplying by 100.

$\frac{9}{25} = \frac{900}{25}\% = 36\%$; $0.023 = 0.023 \times 100\% = 2.3\%$; $\frac{7}{12} = \frac{700}{12}\% = 58\frac{1}{3}\%$

Example 1: Find $14\frac{1}{2}\%$ of £80.

1% of £80 = $\frac{1}{100}$ of £80 = £0.80

14% of £80 = $14 \times$ £0.80 = £11.20

$\frac{1}{2}\%$ of £80 = $\frac{1}{2}$ of £0.80 = £0.40

$14\frac{1}{2}\%$ of £80 = £11.20 + £0.40 = £11.60

Example 2: Express £4.95 as a percentage of £22.50.

$$\frac{£4.95}{£22.50} = \frac{495\text{ p}}{2250\text{ p}} = \frac{11}{50} = \frac{1100}{50}\% = 22\%$$

Example 3: When a man's wage was increased by 15%, he received £12 extra. Find his original wage.

15% of his original wage is £12

1% of his original wage is £$\frac{12}{15}$

100% of his original wage is £$\dfrac{12 \times 100}{15}$ = £80

It is useful to know the following:

$100\% = 1$, $50\% = \frac{1}{2}$, $25\% = \frac{1}{4}$, $75\% = \frac{3}{4}$, $10\% = \frac{1}{10}$, $33\frac{1}{3}\% = \frac{1}{3}$

Exercise 25

1. Express as fractions in their lowest terms: 18%, 35%, 90%, 59%, 44%
2. Express as fractions in their lowest terms: 26%, 30%, 85%, 67%, 8%
3. Express as percentages: $\frac{3}{20}$, $\frac{7}{10}$, $\frac{4}{25}$, $\frac{3}{5}$, $\frac{17}{50}$
4. Express as percentages: $\frac{9}{10}$, $\frac{2}{5}$, $\frac{11}{20}$, $\frac{16}{25}$, $\frac{3}{50}$

5. Express as fractions in their lowest terms: $7\frac{1}{2}\%$, $12\frac{1}{2}\%$, $3\frac{1}{3}\%$, 8.5% $8\frac{1}{3}\%$
6. Express as fractions in their lowest terms: $4\frac{1}{2}\%$, $17\frac{1}{2}\%$, $6\frac{1}{4}\%$, 26.5%, $11\frac{2}{3}\%$
7. Express as percentages: $\frac{1}{8}$, $\frac{3}{8}$, $\frac{1}{6}$, $\frac{3}{16}$, $\frac{4}{7}$
8. Express as percentages: $\frac{5}{8}$, $\frac{7}{8}$, $\frac{5}{16}$, $\frac{3}{7}$, $\frac{2}{9}$
9. Express as decimals: 73%, 9%, 116%, $32\frac{1}{2}\%$, $3\frac{1}{2}\%$
10. Express as decimals: 52%, 7%, 120%, $17\frac{1}{2}\%$, $6\frac{1}{2}\%$
11. Express as percentages: 0.27, 0.06, 1.8, 0.825, 0.0525
12. Express as percentages: 0.42, 0.18, 0.09, 2.7, 0.095
13. Express as percentages, correct to 1 dec. pl.: $\frac{5}{6}$, $\frac{3}{7}$, $\frac{7}{12}$
14. Express as percentages, correct to 1 dec. pl.: $\frac{4}{7}$, $\frac{6}{11}$, $\frac{4}{9}$

Find the value of:

15. 6% of £100	16. 23% of 100 g	17. 87% of £1
18. 50% of £12	19. 10% of 60p	20. 25% of £8
21. 50% of £246	22. 10% of £700	23. 25% of £20
24. $33\frac{1}{3}\%$ of £60	25. 75% of £40	26. 20% of £55
27. 200% of £14	28. 65% of 1 kg	29. 3% of 200 g
30. 45% of £80	31. 34% of £45	32. 7% of £13
33. $17\frac{1}{2}\%$ of £90	34. $2\frac{1}{2}\%$ of £8	35. $6\frac{1}{2}\%$ of £12
36. $4\frac{1}{2}\%$ of £26		

Express the first quantity as a percentage of the second:

37. £7, £100	38. £44, £200	39. £15, £500
40. 30p, £1	41. £7, £25	42. £9, £45
43. £28, £35	44. £6, £24	45. £2.80, £4
46. £2.10, £3.75	47. 27 min, 2 h	48. 650 g, 4 kg

49. A pupil gets 54 marks out of 60. What percentage is this?
50. In a school of 960 pupils, 84 are absent. What percentage is this?
51. There are 120 cars in a car park. 15% of them are red. How many cars are red?
52. A man earning £120 per week receives a 7% rise. Find his new wage.
53. The price of a bicycle was £65. It was increased by 8%. Find the new price.
54. It is said that a man needs 450 grams of food each day. Of this, 14% should be proteins. How many grams of proteins does a man need?
55. A weekly wage is increased from £96 to £108. Express the increase as a percentage of the wage before the increase.
56. The population of a town was 15 000 in 1970. It is now 21 000. Express the increase as a percentage of the 1970 figure.
57. A car was bought for £3200. After one year it was worth only £2640. Express the fall in value as a percentage of the original price.
58. A club had 350 members last year. This year it has 329. Express the decrease as a percentage of last year's membership.

Discount

Example 1: During a sale a shop allows a discount of 10% on the marked prices of all goods. Find the sale price of a coat marked at £34.

The discount is 10% of £34 = £3.40
Sale price = marked price – discount
= £34 – £3.40 = £30.60

Example 2: A store allows a discount of 5p in the £. Find the selling price of an article marked £24.

Discount = 24 × 5p = £1.20
Selling price = £24 – £1.20 = £22.80

Profit per cent

Example 1: A dealer bought a car for £1650 and sold it for £2046. Express his profit as a percentage of (a) his cost price (b) his selling price.

Profit = selling price – cost price = £2046 – £1650 = £396

(a) Profit as % of C.P. = $\frac{396}{1650} \times \frac{100}{1}\% = 24\%$

(b) Profit as % of S.P. = $\frac{396}{2046} \times \frac{100}{1}\% = 19.4\%$ (to 1 dec. pl.)

Exercise 26

1. A shop allows a discount of 10% for cash. How much is paid for a suite of furniture marked £380?
2. A manufacturer allows a retailer a trade discount of 30% of the catalogue price. What does the retailer pay for a TV set listed in the catalogue at £280?
3. A discount of 5p in the £ is allowed on marked prices. What is paid for goods marked (a) £1 (b) £10 (c) £23?
4. On Mondays a grocer gives a discount of 10p in the £. A housewife's bill comes to £3.60 before the discount is taken off. How much does she pay?

In questions **5** to **14**, C P means Cost Price, S P means Selling Price, Pr means Profit and L means Loss. The profit or loss per cent should be based on the cost price.

5. C P = £24, S P = £42. Find Pr and Pr%.
6. C P = £40, S P = £52. Find Pr and Pr%.
7. C P = £35, S P = £21. Find L and L%.
8. C P = £90, S P = £63. Find L and L%.
9. C P = £200, Pr = £50. Find S P and P%.
10. C P = £150, Pr = £18. Find S P and P%.
11. S P = £20, Pr = £4. Find C P and P%.
12. S P = £54, Pr = £9. Find C P and P%.

13. C P = £340, Pr% = 15%. Find Pr and S P.
14. C P = £90, L% = 20%. Find L and S P.
15. An article costing £7.50 is sold for £12.50. Express the profit as a percentage of the selling price.
16. An article costing £3.90 is sold for £6. Express the profit as a percentage of the selling price.
17. When a shopkeeper sold an article for £35, he made a profit of 40% of the selling price. What was the cost price?
18. By selling an article for £80, a dealer made a loss of 25% of the selling price. What was the cost price?
19. An article is sold for £17.50 thereby making a profit of £3.50. (a) Calculate the cost price. (b) Express the profit as a percentage of the cost price.
20. A profit of 15p is made when an article is sold for 55p. Express the profit as a percentage of the cost price.
21. A greengrocer buys a box of 200 oranges for £10 and sells them at 6p each. Find his profit per cent on outlay.
22. Articles costing £8 per 100 are sold at 10p each. Express the profit as a percentage of (a) the cost price (b) the selling price.

Example 1: 24% of a sum of money is £66. Find the sum of money.

24% of the sum is £66

$$1\% \text{ of the sum is } \frac{£66}{24}$$

$$100\% \text{ of the sum is } \frac{£66 \times 100}{24} = £275$$

Example 2: During a sale a shopkeeper allows a discount of 15% on the marked price of a radio. A customer pays £29.75 for the radio. What was the marked price?
As the discount is 15% of the M.P., the customer pays 85% of the M.P.

85% of the M.P. = £29.75

$$1\% \text{ of the M.P.} = \frac{£29.75}{85}$$

$$100\% \text{ of the M.P.} = \frac{£29.75 \times 100}{85} = £\frac{2975}{85} = £35$$

The marked price was £35.

Exercise 27

1. 10% of a sum of money is £6. Find the sum of money. (Remember 10% = $\frac{1}{10}$)
2. 25% of a sum of money is £9. Find the sum of money.
3. 7% of a sum of money is £126. What is 1%? What is the sum of money?

4. 6% of a sum of money is £324. Find 1% and then the sum of money?
5. 40% of a sum of money is £24. What is 10%? What is the sum?
6. 70% of a sum of money is £59.50. Find 10% and the whole sum.
7. 16% of a sum of money is £512. What is the sum of money?
8. 34% of a sum of money is £32.30. Find the sum of money.
9. At a furniture shop the customer was given a discount of 20% of the marked price of a settee. What percentage of the marked price did he pay? If he paid £120, what was the marked price?
10. There is a discount of 30% on the marked prices of all goods in a shop. What percentage of the marked price does a customer pay? £28 is paid for a chair. What was the marked price?
11. During a Bargain Week a shopkeeper gave a discount of 10% on all marked prices. A customer paid £27 for a coat. What was the marked price?
12. A customer paid £357 for a TV set after a discount of 15% had been given on the original price. What was the original price?
13. A discount of 5p in the £ is allowed on marked prices. What was the marked price of an article for which the customer paid (a) 95p (b) £2.85 (c) £15.77?
14. A discount of 12p in the £ is allowed on all marked prices. Find the marked prices of goods for which 88p, £3.52 and £6.38 were paid.
15. 120% of a sum of money is £42. Find 10% of the sum and the whole sum.
16. 115% of a sum of money is £138. Find 5% and the whole sum.
17. When a shopkeeper sold a bicycle for £52, his profit was 30% of his cost price. Hence his selling price (£52) was 130% of his cost price (100% + 30%). From this, calculate the cost price.
18. When selling a radio for £60, a shopkeeper made a profit of 25% of his cost price. Find his cost price.
19. When selling an article for £17.50 a shopkeeper makes a profit of 40% of his cost price. Find his cost price.
20. When selling his car for £468 a man lost 35% of what he paid for it. What did he pay for it?

Simple interest, compound interest and depreciation

Simple interest

A principal of £350 is invested for 4 years at a rate of 6% per annum.

The interest each year is 6% of £350 = £ $\frac{350}{1} \times \frac{6}{100} = £\frac{350 \times 6}{100}$

The interest over 4 years is $£\dfrac{350 \times 6 \times 4}{100} = £84$

If a principal of £P is invested at a rate of R% p.a. for a time of T years, the interest, £I is given by $I = \dfrac{PRT}{100}$

Example: What sum of money, invested for 4 years at 5% p.a., gives £152 interest?

$I = 152$, $R = 5$ and $T = 4$. We have to calculate P.

As $\dfrac{PRT}{100} = I$, $\dfrac{P \times 5 \times 4}{100} = 152$

$$P = \frac{152 \times 100}{5 \times 4} = 760$$

The principal is £760.

Compound interest

At the end of each year (or other period of time) the interest is added to the principal and the interest for the next year is calculated on this new larger principal.

Example: £750 is invested for 3 years at 8% p.a. compound interest.

1st year	Principal	£750	
	Interest	£ 60	(8% of £750)
2nd year	Principal	£810	
	Interest	£ 64.80	(8% of £810)
3rd year	Principal	£874.80	
	Interest	£ 69.984	(8% of £874.80)
	Final amount	£944.784	

The final amount is £944.78, to the nearest penny.

Exercise 28

Find the Simple Interest on:

1. £300 invested for 2 years at a rate of 10% p.a.
2. £500 invested for 4 years at a rate of 8% p.a.
3. £800 invested for 5 years at a rate of 7% p.a.
4. £260 invested for 3 years at a rate of 9% p.a.
5. £160 invested for 5 years at $8\frac{1}{4}$% p.a.
6. £300 invested for 4 years at $9\frac{3}{4}$% p.a.
7. The interest on £360 over 6 years is £108. Find the rate.
8. The interest on £400 at 8% p.a. is £160. Find the time.
9. The interest on £80 over 3 years is £18. Find the rate.
10. What principal is needed so that the interest will be £180 if it is invested at 9% p.a. for 4 years?

11. What principal is needed so that the interest will be £90 if it is invested at 5% p.a. for 3 years?
12. Calculate the simple interest on £642 at 9% p.a. over 2 years.
13. Calculate the simple interest on £73.60 at 10% p.a. over 6 years.

Find the amount obtained in the following cases, the interest being compound and added yearly:

14. £300 invested at 10% p.a. for 2 years.
15. £750 invested at 8% p.a. for 2 years.
16. £800 invested at 5% p.a. for 3 years.
17. £2340 invested at 10% p.a. for 3 years.
18. On 31st December £600 was invested in a building society account when the interest rate was 8% p.a. Each 1st January and 1st July 4% was added for the 6 months period. How much was in the account after one complete year?
19. A man thinks that the value of his car falls each year by 10% of the value at the beginning of the year. His car was worth £1800 last January. What will it be worth (a) next January (b) the following January?
20. The value of some machinery in a factory decreases each year by 20% of its value at the beginning of the year. The machinery cost £7000 when new. What was its value after 2 years?
21. A company borrowed £20 000. At the end of each year it repaid £5000 of which part was interest on the loan at 10% and part was partial repayment of the loan. How much was still owing after 2 years?

Wages and taxes

Exercise 29

Questions **1** to **6** are about wages.

1. A man works 40 hours a week at a rate of £1.80 per hour. What is his weekly wage?
2. A housewife takes a part-time job for which she is paid £2.40 per hour. She works 3 hours each day from Monday to Friday. What is her weekly wage?
3. For a certain job the salary for the first year is £3380. Each year the salary is increased by £260. Thus for the second year it is £3640. (a) Calculate the salary for the fourth year. (b) In which year is the salary £5200? (c) In the first year, how much is received per week?

4. A man is paid at the basic rate of £3 per hour. If he works more than 36 hours in a week, the extra time is paid for at $1\frac{1}{2}$ times the basic rate. One week he worked for 42 hours. What was his wage for the week?
   Another week his wage was £117. How much overtime did he do?
5. A salesman receives £180 per month plus a commission of 2% of the value of the goods he sells. One month he sells goods to the value of £2900. How much is he paid for the month?
6. A traveller has a salary of £3000 per year plus a commission of 3% on all sales. His sales for a year were £80 000. What was his total income?

Questions **7** to **16** are about taxes.

7. A person's taxable income is £1800. If the rate of tax is 30%, calculate the tax to be paid.
8. A man earned £5330 one year. He paid no tax on the first £1860 and he paid 30% tax on the remainder. How much tax did he pay?
9. When income tax was 30% a person paid £120 in tax. What was his taxable income?
10. VAT was levied at 8%. How much VAT was added to a bill for £55.
11. A calculator was advertised for £14.40 plus VAT at 15%.
    Find the cost after adding VAT.
12. A camera was sold for £23 which included VAT at 15%. Find the price before VAT was added.
13. A television set was sold for £368 which included VAT at 15%. Find the price before VAT was added.
14. In shop A a cassette recorder was marked £44 including 10% VAT. In shop B the same model was marked £41 plus 10% VAT. Which was the cheaper and by how much?
15. A man's salary was £4800 per annum. From this was deducted 6% p.a. for a pension scheme and £22.70 per month for National Insurance. His income tax was £42 per month. How much was left each month after these deductions?
16. One year a man earned £4580. He paid no tax on the first £1940. On the next £750 he paid tax at 25% and on the remainder he paid tax at 30%. How much tax did he pay?

Insurance and hire purchase

Exercise 30

Questions **1** to **6** are about insurance.

1. A person wishes to take out a life assurance policy for £2000 payable in 15 years time. The monthly premium is quoted as £0.58 per £100. How much will he have to pay (a) per month (b) per year?
2. An annual premium of £4.42 per £100 is quoted for a 25 years life assurance policy. A man wishes to take out a policy for £3000. How much will he have to pay (a) per year (b) per week. (Take 52 weeks to a year.)
3. A man insured his house for £16 000 and its contents for £4000. The annual premium for the house was 9.5p per £100 and for contents was 22p per £100. Calculate the total premium.
4. A man values his house at £14 000 and its contents at £3600. Find his annual insurance premium if the rate for the building is £1.10 per £1000 and the rate for the contents is £2.80 per £1000.
5. A man pays an annual premium of £16.20 on the contents of his house, the rate being 45p per £100. What value does he put on the contents?
6. A man took out a life assurance for £4000. The monthly premium was £0.42 per £100. How much did he pay per year? There was income tax relief at the rate of 30% on half the premium paid. Calculate his tax relief.

Questions **7** to **14** are about hire purchase.

7. A bicycle can be bought for a deposit of £23 and 20 monthly payments of £1.80. What is the total amount to be paid?
8. The HP terms for a certain cassette player are a deposit of £14.90 and 12 monthly payments of £2.20. What is the total amount to be paid?
9. The cash price of a sewing machine is £35. It can be bought by paying a deposit of 20% of the price and paying the rest in 10 weekly instalments. How much must be paid per week?
10. A radio can be purchased for £39.95 cash or by a deposit of £10 and 12 monthly payments of £2.75. How much more is paid by the second method?
11. Some furniture was advertised at £235 cash. The HP terms were £20 deposit and 24 monthly payments of £9.30. Calculate (a) the total HP price (b) the difference between the cash price and the HP price.
12. The price of a cooker is £84. A person pays a deposit of £18. The remaining amount has an interest charge of 8% added and is paid by 12 monthly instalments. Calculate (a) the interest (b) the monthly instalment.
13. A refrigerator costs £72. It can be paid for by a deposit of 25% of the

price and the remainder, which has an interest charge of 10% added, is paid in 12 monthly instalments. Calculate (a) the deposit (b) the interest (c) the monthly instalment.

14. The cash price of a camera is £60. If you make an initial payment of 20%, how much should you pay each month to clear the balance in twelve months, if no interest is charged?
    If you were asked to pay £4.20 per month, how much extra would you pay over the twelve months? Express this extra as a percentage of the balance.

Rates, rent and mortgages

Exercise 31

Questions **1** to **12** are about rates.

1. Calculate the amount to be paid in rates on a house assessed at £150 if the rate is £0.60 in the £.
2. The rateable value of a shop is £420. Calculate the amount to be paid if the rate is 82p in the £.
3. The rateable value of a house is £120 and the owner pays £84 in rates. Calculate the rate in the £.
4. The rateable value of a house is £220 and a rate demand for £143 is received by the owner. Calculate the rate in the £.
5. A householder pays £132 in rates when the rate is 55p in the £. Calculate the rateable value of the house.
6. A rate demand for £126 is received when the rate is £0.70 in the £. Calculate the rateable value of the property.
7. The rateable value of the property in a town is £6 327 000 and a sum of £4 984 000 is required by the council. What rate, to the nearest penny, must be charged?
8. The rateable value of a district is £1 500 000. The council needs to raise at least £460 000 by levying a rate that is a whole number of pence per pound. Calculate the least rate that can be levied.
9. When the rate was 70p in the £, a householder paid £126 in rates. The rate was increased by 6p in the £. How much more did he have to pay?
10. The rateable value of a town is £8 460 000. If the rate is increased by 2p in the £, what additional revenue will this bring?
11. The rateable value of a house is £280.
    (a) Calculate the rates payable when the rate is 70p in the £.
    (b) In a later year the sum to be paid is £238. Calculate the new rate in the £.

12. In a certain district each 1p in the £ was estimated to bring £53 000. What was the rateable value of the property in the district? For the Fire Service a rate of 2.55p in the £ was needed. What was the estimated cost of the Fire Service?

Questions **13** to **17** are about rent.

13. A landlord charged a rent of £6.50 per week for a flat. How much did he receive per year?
14. The weekly rent for a house is £8.20. How much does the tenant pay per year?
15. A man owns a house which is divided into five bed-sitters. He lets two of them at £4.50 each per week and the others at £5.50 each per week. How much does he receive (a) per week (b) per year?
16. A television set can be rented for £2.20 per week. The same set can be bought for £290. After how many weeks will the rent paid be more than the price of £290?
17. A flat was rented at £5.50 per week plus rates. The rateable value was £130 and the rates were 60p in the £. Calculate (a) the rates for a year (b) the rates per week (c) the weekly rent after adding the rates.

Questions **18** to **21** are about mortgages.

18. A man takes out a mortgage for 25 years. He pays £936 each year. (a) How much is this per week? (b) What is the total amount paid over the 25 years?
19. A certain building society will not give a mortgage of more than $2\frac{1}{4}$ times a person's annual income. Mr Fraser has an income of £4800. What is the largest mortgage he can obtain?
20. A man buying a house needs a mortgage for £12 000 and is quoted £8.20 per £1000 per month. Calculate his payment (a) per month (b) per year.
21. A man borrows £10 000 when buying a house. He agrees to repay £420 of the loan each year together with interest at the rate of 12% per annum on the amount of the loan outstanding at the beginning of the year.
    (a) How much does he pay for the first year?
    (b) How much does he pay for the second year?
    (c) How much does he pay for the year when £4000 of the loan has been repaid?

Gas and electricity

Exercise 32

1. A householder was charged for electricity as follows:
   5p per unit for the first 80 units and 1.4p per unit for the rest. If he used 420 units in a quarter, how much was his bill?
2. A householder used 314 units of electricity in a certain quarter. He has to pay a fixed charge of £2.80 and 2.5p for each unit used. Calculate his bill for the quarter.
3. The readings on a gas meter at the beginning and end of a quarter were 3824 and 3956 units. The price per unit was 22p and there was a standing charge of £2. Calculate the gas bill.
4. For gas a householder paid 22.8p per therm for the first 50 therms used and 15.3p per therm for the rest. There was also a quarterly standing charge of £2.10. If 140 therms were used one quarter, what was the bill?
5. A club pays for electricity under a tariff which consists of a charge of 1.32p for each unit used and a quarterly charge of £6.40. Calculate the bill for a quarter in which 426 units were used. Give your answer to the nearest penny.
6. At one time the electricity tariff for industry was 6.2p per unit for the first 100 units, 4.2p per unit for the next 250 units and 3.0p per unit for the rest. Calculate the bill for a workshop which used 860 units.
7. In a certain area there was a choice of the following two tariffs for gas. Tariff A: 14p per therm for the first 20 therms and 10p per therm for the rest. Tariff B: a fixed charge of £1.50 and 6p per therm. Find the cost on each tariff for using 36 therms.
8. The readings on an electricity meter were 52 847 on August 3 and 54 162 on November 6. The cost of each unit was 2.746p and there was a standing charge of £2.84. Calculate (a) the number of units used (b) the total charge, to the nearest penny, for the quarter.

Foreign currency

Exercise 33

1. When £1 was equal to 8.40 French francs, how many francs were obtained for £3.50?
2. If £1 = 142 Spanish pesetas, how many pesetas are obtained for £16.50?

3. How many German marks should I get for two £10 and one £5 travellers cheques if £1 = 3.68 marks?
4. Given that £1 = 8.40 French francs and 1 franc = 195 Italian lire, calculate the number of lire to £1.
5. If £1 = 3.68 marks, calculate, to the nearest penny, the value in British money of 1 mark.
6. If £1 = 1.95 U S dollars, how much British money, to the nearest penny, can be obtained for 20 dollars?
7. In a Swiss shop a watch was marked 84 francs. Change the price to British money given that £1 = 3.2 Swiss francs.
8. If £1 = 1.95 U S dollars and £1 = 390 Japanese yen, how many dollars should be obtained for 6000 yen?
9. If £1 = 4 Dutch guilders and £1 = 58 Belgian francs, how many francs would be obtained for one guilder?
10. A holiday maker in Italy cashed travellers' cheques for £60 at the rate of £1 = 1612 lire. He spent 75 430 lire. On returning home he changed the rest into British money at the rate of £1 = 1654 lire. How much British money did he get?
11. A traveller in Germany changed £50 into marks at the rate of £1 = 3.62 marks. He spent 135 marks and changed the rest back into British money at the rate of £1 = 3.68 marks. How much did he get?
12. A meal for four people in France cost 58.80 francs. If £1 = 8.40 francs, calculate the cost per person in British money.
13. A pupil has the following Swiss coins: 3 of 5 fr, 4 of 1 fr, 5 of 50 c, 7 of 20 c and 3 of 10 c. 1 fr (franc) = 100 c (centimes). What is the total value of the coins (a) in Swiss francs (b) in British money if £1 = 3.2 francs?

Bar charts, pie charts and line graphs

In a pie chart (Fig. 1 and 2), each number is represented by the area of a sector of a circle and so by the angle of the sector.
In a bar chart (Fig. 3) each number is represented by the height of a bar, the bars all having the same width.
For a line graph we mark points to represent the data and then join the points with lines to show upward or downward trends.

Exercise 34

1.

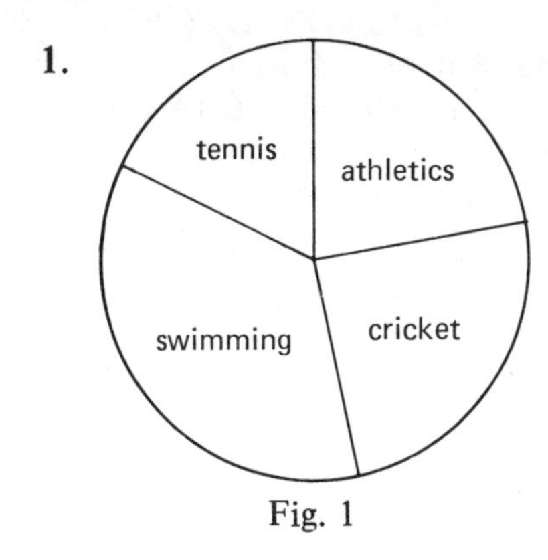

Fig. 1

The pie chart in Fig. 1 shows how 90 pupils were divided for various activities in a games period. Which activity had the most pupils? Which had the least?

As 360° represents 90 pupils, 4° represents 1 pupil. The angle used for swimming is 128°. $128^\circ \div 4^\circ = 32$ and so 32 pupils went swimming. The angles for athletics, cricket and tennis are 80°, 88° and 64°. How many pupils took part in each of these activities?

2.

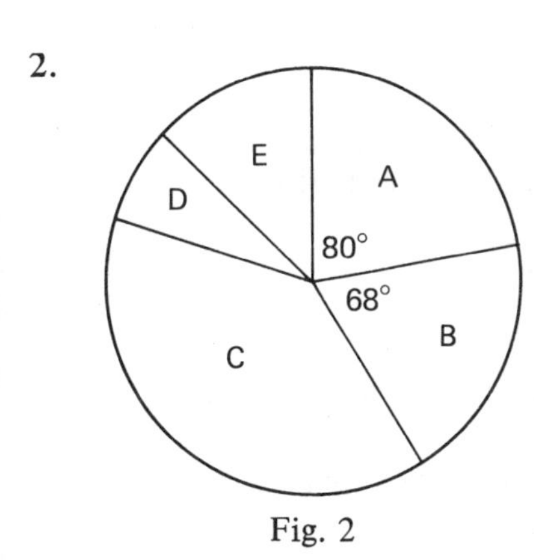

Fig. 2

The pie chart in Fig. 2 represents 180 pupils. (a) What angle represents 1 pupil? (b) How many pupils are represented by sector A and how many by sector B? (c) If sector C represents 70 pupils and sector D represents 13 pupils, what angles do they have? (d) What is the angle of sector E and how many pupils does it represent?

3. Of 120 pupils who left school one term, 30 started work in factories, 20 started in offices, 10 started in shops, 40 had places in universities and colleges and 5 were going to other schools. How many does this leave out of the 120? When drawing a pie chart for the 120, what angle should be used for 10 pupils? State the angle to be used for each sector and draw the pie chart.
4. A class has 36 lessons each week. They are allotted as follows: English 4, Mathematics 5, Science 6, French 4, Geography 3, History 3 and others 11. Draw a pie chart for this data.
5. A boy stated that he had spent the 24 hours of one day as follows: at school 7 h, watching TV 2 h, asleep 8 h, playing football $1\frac{1}{3}$ h, other activities $5\frac{2}{3}$ h. For a pie chart, what angle should represent 1 hour? State the angle needed for each of the sectors and draw a pie chart.

6. The bar chart in Fig. 3 shows how the fifth year pupils travel to a certain school. (a) What is the commonest method? (b) What is the least common method? (c) How many pupils walk to school? (d) What means of transport is used

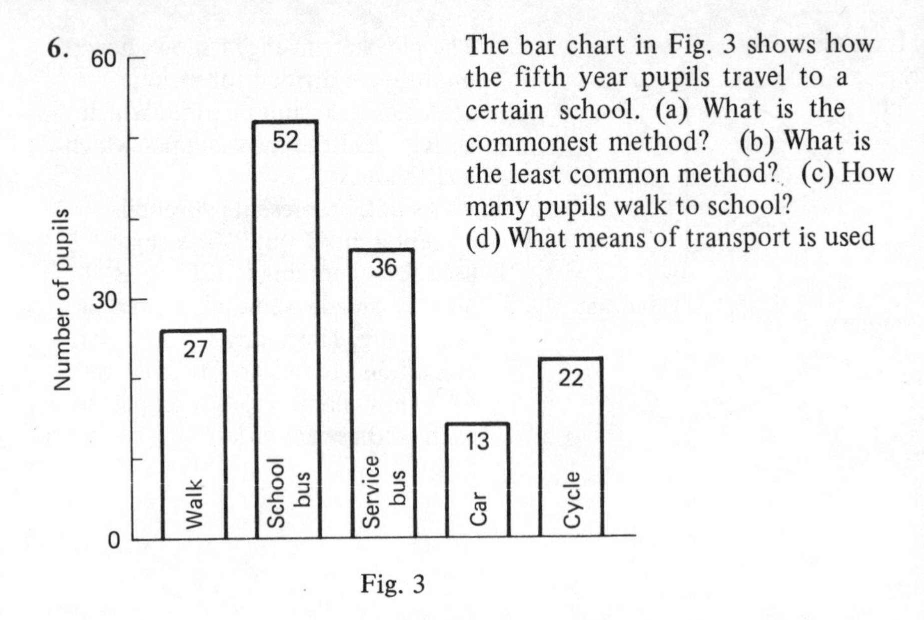

Fig. 3

by 22 pupils? (e) How many more come by school bus than by service bus? (f) How many pupils are there in the fifth year? (g) What percentage walk to school?

7. The number of coats sold in a shop each day of a certain week was as follows: Monday 8, Tuesday 10, Wednesday 7, Thursday 4, Friday 12, Saturday 23.
Draw a bar chart for the data using 5 mm to 1 coat and making each bar 10 mm wide. Which was early closing day?

8. The table gives the number of calculators sold by a shop in each of five years:

1974	1975	1976	1977	1978
160	270	420	680	750

Represent the data by a bar chart, using 1 mm to 10 calculators.

9. Out of every 100 people who went on holiday in the U.K. in 1976, 27 went to hotels or guest houses, 20 went to caravans, 25 stayed with friends or relations, 11 rented houses or flats and 7 went camping. How many spent their holidays in other ways? Draw a bar graph using 2 mm for one person.

10. The heights of a person on certain birthdays were as follows:

Age in years	10	12	14	16	18	20
Height in cm	124	132	148	162	170	170

For a graph of this data, take a horizontal axis from 10 years to 20 years using 1 cm to 1 year and a vertical axis from 0 to 180 cm using 1 cm to 10 cm. Plot points for the data and join them.
(a) Between what ages did the person not grow at all? (b) Between

what ages did he grow most? (c) Estimate his height at 15 years of age. (d) Estimate his age when he was 128 cm high.

11. The table shows the temperature in a room at various times one day.

Time	9.00	10.00	11.00	12.00	13.00	14.00	15.00
Temp. (°C)	15.0	15.5	16.8	17.8	18.8	19.0	18.2

Time	16.00	17.00	18.00
Temp. (°C)	17.4	17.0	16.5

Using 1 cm for 1 hour and 2 cm for 1°C, plot points to represent the data and join them with a smooth curve. From your graph estimate (a) the temperature at 11.30 (b) the times when the temperature was 18.0°C.

Mean, median and mode

In nine tests a pupil obtained the following marks:

8, 4, 2, 4, 9, 5, 10, 4, 8

The **mean** of the marks is $\dfrac{\text{the sum of the marks}}{\text{the number of tests}} = \frac{54}{9} = 6$

Arranging the marks in order of size, we have

2, 4, 4, 4, 5, 8, 8, 9, 10

The most frequent mark, 4, is called the **mode.**
The middle mark, 5, is called the **median.**
For the median of an even number of tests, we select the middle two marks and find their mean. For example, the median of
12, 13, 15, 18, 20, 23 is $\frac{1}{2}(15 + 18) = 16\frac{1}{2}$.

Example: A class of 30 pupils has 18 boys and 12 girls. The mean mass of the boys is 71 kg and the mean mass of the girls is 59 kg. Calculate (a) the total mass of the boys (b) the total mass of the girls (c) the mean mass of the 30 pupils.

The total mass of the boys = 18 × 71 kg = 1278 kg
The total mass of the girls = 12 × 59 kg = 708 kg

The total mass of the 30 pupils = 1278 kg + 708 kg = 1986 kg
The mean mass of the pupils = 1986 kg ÷ 30 = 66.2 kg

Exercise 35

Questions on means:

1. Find the means of the following sets of numbers:
   (a) 4, 6, 8, 8, 9 (b) 7, 8, 11, 15, 17, 20
   (c) 0.3, 0.6, 0.7, 0.9, 1.0, 1.0, 1.1
2. A batsman made the following scores in eight innings: 15, 33, 1, 27, 18, 49, 17, 0. Find his mean score per innings.
3. The number of pages in a daily newspaper for six days were 24, 32, 28, 32, 28, 36. Calculate the mean number of pages per day.
4. The mean of four numbers is 18. What is the sum of the four numbers? Three of them are 20, 24 and 15. What is the other?
5. For a certain week the mean takings at a shop were £58 per day. What were the total takings for the week, the shop being open for six days?
6. The ages of five children in a family are 5, 7, 10, 11, 12 years. Find the mean age (a) now (b) 4 years ago (c) in 10 years time.
7. The heights of six pupils are 154, 156, 162, 151, 160, 159 cm. Make a list of the amounts by which the heights exceed 150 cm. Find the mean of these excesses and add on 150 cm to obtain the mean height of the pupils.
8. The ages at which eight men retired were 60, 65, 62, 64, 67, 60, 65, 65. Calculate the mean age. (Take 60 off each first.)
9. The mean mass of a boat crew of eight men is 69.2 kg. Find the total mass. A man of mass 67.8 kg is replaced by one of mass 72.6 kg. Find the new mean mass.

Questions on medians:

10. State the median (middle value) of each of the following sets of numbers: (a) 2, 5, 7, 9, 10 (b) 26, 30, 35, 40, 44, 48, 53
    (c) 1, 4, 6, 9 (d) 53, 58, 60, 65, 65, 70
11. Arrange the following in order of size and then state the median.
    (a) 6, 3, 7, 2, 5 (b) 14, 13, 9, 11, 6 (c) 7, 0, 4, 3, 2, 8
    (d) 0.3, 0.4, 0.2, 0.3, 0.2, 0.3
12. The masses of eight people are 46, 41, 49, 40, 43, 52, 51, 42 kg.
    (a) Find the median mass. (b) Calculate the mean mass.
13. On a cycling tour, a boy travelled the following daily distances: 104, 84, 117, 50, 95 km. (a) Find the median distance
    (b) Calculate the mean distance.

Questions on modes:

14. State the mode of each of the following:
    (a) 3, 5, 6, 8, 8, 8, 9 (b) 2, 2, 3, 4, 4, 4, 5, 5
15. State the mode of each of the following:
    (a) 3, 7, 6, 3, 3, 7 (b) 4, 2, 2, 0, 6, 1, 2, 3, 0, 4
16. A football team scored the following number of goals in ten matches:

3, 1, 2, 1, 0, 2, 2, 1, 5, 2. State the mode. Also calculate the mean.

17. The marks of fifteen pupils in a test were: 7, 9, 9, 6, 8, 7, 10, 9, 7, 6, 8, 7, 7, 4, 7. Find the mode and the mean.

18. The shoe sizes of ten children are 7, 4, 7, 7, 5, 7, 4, 8, 5, 6. Find the mode and the mean.

19. In a competition nine teams were awarded points as follows: 6, 11, 11, 11, 12, 14, 14, 15, 16. State the mode and the median. Calculate the mean, correct to one decimal place.

Frequency distributions

Histograms

This is a list of the number of goals scored by 40 football teams one Saturday:

4 0 1 2 1 3 0 2 1 0 1 1 6 4 1 2 2 2 4 6
0 0 2 3 1 1 2 2 4 1 3 0 2 1 2 0 3 3 1 1

From the list we can form the following tally chart and frequency table.

Number of goals	Tally	Frequency
0	~~1111~~ 11	7
1	~~1111~~ ~~1111~~ 11	12
2	~~1111~~ ~~1111~~	10
3	~~1111~~	5
4	1111	4
5		0
6	11	2

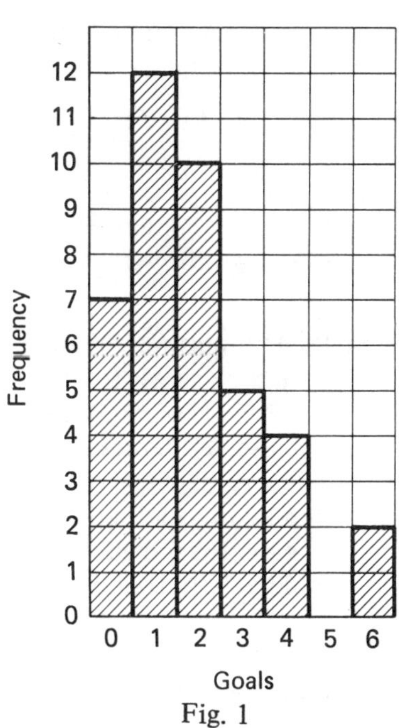

Fig. 1

The table shows that 7 teams scored no goals, 12 teams scored 1 goal, etc. We say that the score of 0 occurred with a frequency of 7, the score of 1 occurred with a frequency of 12, etc.

Fig. 1 shows a **histogram** representing the information. Each square represents 1 team.

Exercise 36

1. (a) From the above frequency table, how many teams scored 3 goals?
   (b) How many teams scored less than 3 goals?
   (c) How many teams scored more than 3 goals?
   (d) What was the mode?
2. 30 pupils obtained the following grades in an Arithmetic examination.
   3 2 4 3 2 1 3 3 5 4 4 2 3 3 4
   4 5 3 5 4 3 4 1 2 3 3 2 5 4 3
   Prepare a tally chart and frequency table as above and draw a histogram. State the mode. How many obtained grade 1 or 2?
3. The following grades were obtained by 40 pupils in an English examination.
   5 4 3 5 1 4 2 3 2 5 4 3 4 6 5 1 5 2 3 3
   4 5 4 6 4 3 5 4 4 2 5 4 6 3 2 4 3 4 3 5
   Prepare a tally chart and frequency table. Draw a histogram and shade the area representing those who obtained grades 1 or 2. Express this number as a percentage of the 40.
4. A gardener counted the number of flowers on each plant in a flower bed. He obtained this frequency table.

Number of flowers	4	5	6	7	8	9	10	11	12
Number of plants	1	3	11	17	18	13	7	2	1

   Draw a histogram. State the mode. State the number of plants examined. How many plants had fewer than seven flowers?

5. Some pupils were asked how long they took to get to school. Fig. 2 shows a histogram for the results. Each square represents 2 pupils. The left rectangle represents 8 pupils who took up to 5 minutes. The next rectangle represents the number who took between 5 and 10 minutes. From the histogram, copy and complete this frequency table.

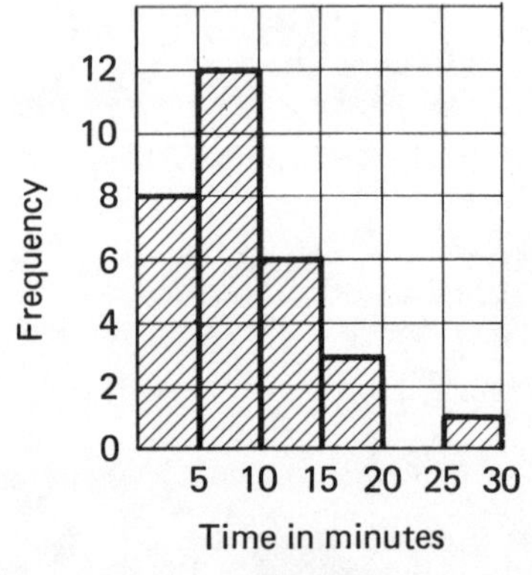

Fig. 2

Time taken (minutes)	0–5	5–10	10–15	15–20	20–25	25–30
Frequency (no. of pupils)	8					

6. Some pupils were asked how much pocket money they received the previous week. Fig. 3 shows the results. Each square represents 1 pupil. From the histogram form a frequency table as in question **5**.
How many pupils were questioned? How many received more than 70p? What was the modal class (most frequent range of pocket money)?

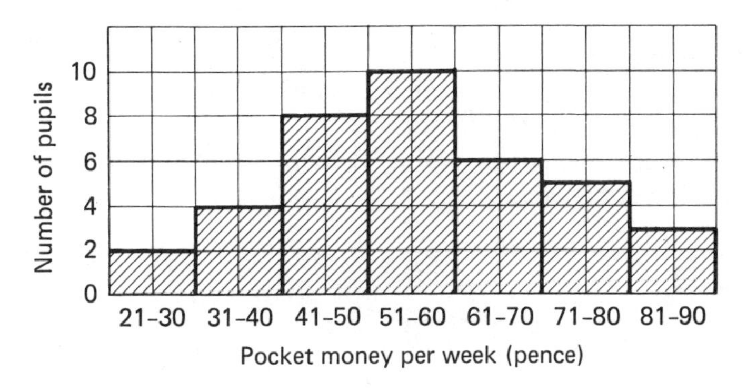

Fig. 3

7. Eighty pupils were asked to state the distances of their homes from their school. The results are shown in the table.

Distance in km	Under 2	2 to 4	4 to 6	6 to 8
Frequency	28	20	24	8

Draw a histogram using 1 cm to 1 km on the distance axis and an area of 1 cm^2 for 1 pupil.
State the modal class. What percentage of pupils lived more than 4 km from the school?

8. Two hundred newly born babies were weighed and the following table was obtained.

Mass (kg)	1.5-2.0	2.0-2.5	2.5-3.0	3.0-3.5	3.5-4.0	4.0-4.5	4.5-5.0
Frequency	5	10	45	60	55	15	10

State the modal class. Draw a histogram using 1 cm for 0.5 kg on the mass axis and an area of 1 cm^2 for 5 babies.

9. The marks for 100 pupils who took an examination were grouped to give the following table.

Marks	11-15	16-20	21-25	26-30	31-35	36-40	41-45	46-50
Frequency	6	11	14	19	23	15	8	4

Draw a histogram for this data. Shade the area representing those who scored more than 40 marks.

10. The masses of 80 pupils were measured to the nearest kilogram and the results grouped to give the following table.

Mass (kg)	41-44	45-48	49-52	53-56	57-60	61-64	65-68	69-72
Frequency	2	7	14	20	18	11	6	2

Draw a histogram for this data. Shade the rectangle representing the modal class. Express the number of pupils in this class as a percentage of the total.

Frequency polygons

A frequency distribution can be illustrated by a frequency polygon. Points are plotted to represent the frequencies and joined by straight lines to show the pattern. Polygons are useful for comparing two or more frequency distributions as in Fig. 4.

Exercise 37

1.

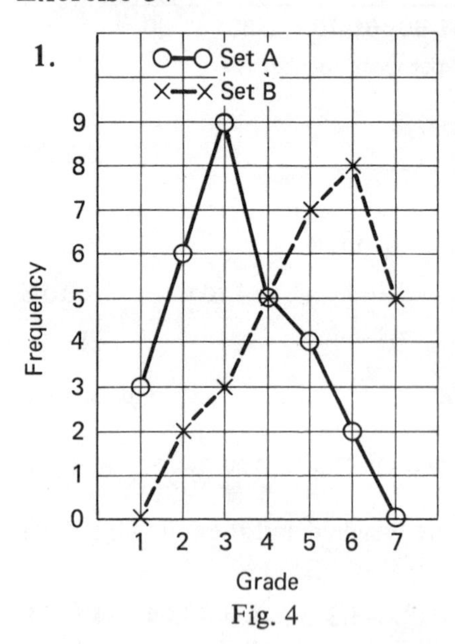

Fig. 4

The pupils in two sets were given grades 1 to 7 in an Arithmetic test. Fig. 4 shows the resulting frequency polygons. We see that 3 pupils in set A obtained Grade 1 but no pupils in set B. Copy and complete the following frequency table:

Grade	1	2	3	4	5	6	7
Frequency, set A	3	6					
Frequency, set B	0	2					

State the mode for each set.

2. In a test taken by 50 pupils, the marks were as follows:

Mark	0	1	2	3	4	5	6	7	8	9	10
Frequency	0	1	4	6	7	8	8	6	5	3	2

Draw a frequency polygon to represent this data.

3. A pupil counted the number of letters in each of the first hundred words of a newspaper article. His results are shown in the following frequency table:

Number of letters per word	1	2	3	4	5	6	7	8	9	10	11	12
Frequency (number of words)	5	17	13	12	15	10	12	7	2	4	2	1

Draw a frequency polygon to represent this data.

4. The table shows the results for a class of 30 pupils in an Arithmetic examination and in an English examination.

Grade	1	2	3	4	5	6	7	8
Frequency: Arithmetic	0	0	5	6	12	5	2	0
Frequency: English	3	4	7	8	4	2	1	1

Draw two frequency polygons to represent this data. Comment on the two distributions.

5. Two football teams each played 26 matches. The goals scored were as follows:

Wanderers	1	0	3	5	0	2	1	0	0	3	1	0	1
	1	2	0	1	1	0	0	1	5	0	2	1	0
Rangers	3	1	4	2	0	3	2	1	1	5	0	1	2
	4	6	2	3	2	5	4	3	2	2	1	2	1

Produce a frequency table as in question **4** and draw two frequency polygons. Comment on the distributions.

6. The percentages of the population of the United Kingdom in various age ranges in 1910 and in 1967 are shown below:

Age range	Under 10	10–20	20–30	30–40	40–50	50–60	60–70	70–80	Over 80
% of pop. in 1901	22	20	18	14	11	7	5	2	1
% of pop. in 1967	17	14	14	12	13	12	10	6	2

Draw two frequency polygons and comment on the distributions. Each point should be plotted at the centre of the class, i.e. at 5, 15, 25, etc.

Means of frequency distributions

20 pupils gave some money to a certain collection. 8 gave 5p each, 9 gave 10p each and 3 gave 50p each.

Altogether they gave $8 \times 5\text{p} + 9 \times 10\text{p} + 3 \times 50\text{p}$
$= 40\text{p} + 90\text{p} + 150\text{p} = 280\text{p}$

The mean sum given $= \dfrac{\text{total sum given}}{\text{number of pupils}} = \dfrac{280\text{p}}{20} = 14\text{p}$

The calculation can be set out in columns as follows:

Money given	Frequency	Total
5p	8	40p
10p	9	90p
50p	3	150p
	20	280p

$$\text{Mean} = \frac{280\text{p}}{20} = 14\text{p}$$

Exercise 38

Show your calculations in columns as above:

1. 30 pupils gave money to a collection. 6 gave 5p each, 15 gave 10p each and 9 gave 20p each. Calculate (a) the total amount given (b) the mean.
2. I bought 7 books costing £1 each, 8 costing £2 each and 5 costing £3 each. (a) What was the total cost? (b) What was the mean cost per book?
3. 20 pupils did an Arithmetic test. 4 obtained 6 marks each, 8 obtained 7 marks each, 6 obtained 8 marks each and 2 obtained 9 marks each. Calculate (a) the total marks received (b) the mean mark.
4. A football team received 3 points for a win, 2 points for a draw and 1 for a loss. They won 11 matches, drew 8 and lost 6. Calculate (a) the total number of matches played (b) the total number of points received (c) the mean number of points per match.
5. The number of peas in each of 50 pods was counted and the following frequency table obtained:

Number of peas in a pod	4	5	6	7	8
Number of pods (frequency)	4	14	18	11	3

   Calculate (a) the total number of peas (b) the mean number of peas per pod.
6. One Saturday afternoon the number of goals scored by 50 teams were as follows: 8 scored 0, 21 scored 1, 12 scored 2, 6 scored 3 and 3 scored 4.
   Calculate (a) the total number of goals scored (b) the mean number of goals per team.
7. The times taken by 15 pupils to run round a cross-country course were recorded to the nearest minute and the following table was obtained:

Time (minutes)	11-13	14-16	17-19	20-22
Frequency (number of pupils)	2	4	6	3

Taking the time for each class to be at the mid-point of the interval (12 for the first class), calculate the mean time taken.

8. The heights of some plants were measured and the following table was obtained:

Height of plant (cm)	16-20	21-25	26-30	31-35
Frequency (no. of plants)	3	6	7	4

(a) What is the middle of the first class interval?
(b) Using the middle heights, calculate the mean height.

Range and interquartile range

Betty's marks in eight tests were: 5, 5, 6, 6, 7, 7, 7, 7
David's marks in the same tests were: 1, 2, 4, 6, 7, 8, 9, 9
Both have the same median mark of $6\frac{1}{2}$, but Betty's marks are all close to $6\frac{1}{2}$ whereas David's are widely spread out. One way of measuring the spread or **dispersion** is by using the range. This is the difference between the highest and lowest marks. For Betty it is $7 - 5 = 2$; for David it is $9 - 1 = 8$.
If Betty's marks had been 2, 5, 6, 6, 7, 7, 7, 7, her range would have been 5. Although seven of her marks would be unchanged, the one bad mark would make the range much greater. To avoid this we often use another method for the spread. The median divides a list of numbers into two equal parts. The **quartiles**, together with the median, divide the list into four equal parts. For David's marks we have:

$$1, 2 \underset{Q}{\uparrow} 4, 6 \underset{M}{\uparrow} 7, 8 \underset{Q}{\uparrow} 9, 9$$

The lower quartile is $\frac{1}{2}(2 + 4) = 3$; the upper quartile is $\frac{1}{2}(8 + 9) = 8\frac{1}{2}$.
The **interquartile range** is (upper quartile − lower quartile) $= 8\frac{1}{2} - 3 = 5\frac{1}{2}$.
For Betty's marks the quartiles are $5\frac{1}{2}$ and 7 and so the interquartile range is $1\frac{1}{2}$.

Exercise 39

1. Arrange each set of numbers in order of size and then state the range:
   (a) 10, 7, 9, 13, 11 (b) $7\frac{1}{2}$, 9, 12, $5\frac{1}{2}$, 10, 7
   (c) 62, 91, 43, 28, 37, 26, 54, 31
2. State the lower quartile, the median, the upper quartile and the interquartile range for each of the following sets of numbers:
   (a) 3, 4, 6, 7, 9, 9, 11, 11 (b) 2, 5, 6, 9, 10, 12, 13, 17
   (c) 11, 11, 12, 14, 17, 20, 24, 28, 32, 38, 40, 46

3. Arrange each set of numbers in order of size. State the lower quartile, the median, the upper quartile and the interquartile range:
(a) 5, 8, 0, 3, 7, 10, 12, 7 (b) 7, 10, 5, 5, 9, 7, 5, 9
(c) 8, 6, 7, 8, 4, 2, 1, 8, 9, 7, 7, 4
4. The marks obtained by three girls in a set of tests were
Alice: 5, 6, 8, 6, 0, 7, 5, 6
Brenda: 8, 9, 6, 4, 2, 1, 1, 9
Carol: 3, 5, 6, 5, 4, 4, 5, 6
Compare the three sets of marks by finding their medians and their interquartile ranges.
5. The hours of sunshine at two towns for eight days were:
Brightsea: 8, 9, 7, 10, 11, 12, 0, 9
Greyley: 0, 2, 10, 3, 6, 0, 0, 3
Compare the sets of figures by (i) their ranges (ii) their interquartile ranges.
6. The average monthly temperatures for two towns are:
Deviton: 6, 11, 13, 17, 20, 22, 28, 25, 22, 20, 12, 8 °C
Constroft: 14, 14, 16, 18, 19, 20, 21, 19, 18, 16, 15, 14 °C
Show that the towns have the same mean temperature. Compare the variations in temperatures by finding the interquartile ranges.

Probability

Ten cards bearing the letters A, A, A, D, T, S, D, P, F, H are shuffled and one is drawn out. As there are three cards with the letter A, the probability of drawing an A is $\frac{3}{10}$. We write $p(A) = \frac{3}{10}$ or 0.3. Similarly $p(D) = \frac{2}{10} = \frac{1}{5}$ or 0.2 and $p(H) = \frac{1}{10}$ or 0.1. If a trial has n equally likely outcomes (such as any of the above ten cards) and a certain event (such as an A) can happen in x ways, the probability of the event is x/n.
If an event is certain to happen, its probability is 1; if an event is impossible its probability is 0.
If the probability of an event happening is p, then the probability of it not happening is $1 - p$. For the above cards $p(\text{not A}) = 1 - 0.3 = 0.7$.
When a die is thrown $p(\text{a six}) = \frac{1}{6}$. If a die is thrown 60 times, the expected number of sixes is $\frac{1}{6}$ of $60 = 10$. (In practice it is very unlikely to be exactly 10 but is likely to be near to 10.)

Exercise 40

1. A die is thrown. State the probability of (a) 5 (b) an odd number (c) a number less than 3 (d) 7.
2. A card is drawn from a pack of 52 playing cards. State the probability of (a) a spade (b) a red card (c) an ace (d) a court card (king, queen or jack).
3. Cards bearing the numbers 1 to 9 are placed in a box and one is drawn out at random. State the probability of (a) an even number (b) an odd number (c) 7, 8 or 9 (d) less than 5.
4. A class of 30 pupils has 12 girls. A name is chosen at random. What is the probability it is a girl's name.
5. A box contains 4 yellow, 1 green and 5 white pieces of chalk. One is drawn out at random. What is the probability it is (a) green (b) yellow (c) white (d) red?
6. A box contains 6 red, 9 blue and 10 black pens. One is taken out at random. What is the probability it is (a) blue (b) red (c) not red (d) not black?
7. A coin is spun 100 times. What is the expected number of heads?
8. A die is thrown 60 times. How many times would you expect to get (a) an odd number (b) a one (c) a one or a two?
9. Assuming that the probability of being born in January is $\frac{1}{12}$, how many pupils in a school of 600 are likely to have birthdays in January?
10. A box contains 30 beads, some blue and some orange. If one is taken out, p(blue) = 0.3 and p(orange) = 0.7. How many are there of each colour?
11. There are 36 cards in a pile. Some are red and the others are black. When one is taken out at random, p(red) = $\frac{2}{9}$. What is p(black)? How many of each kind are there in the pile?
12. A box contains twenty sweets. Four are red, six are yellow and the rest are green. One is taken out at random. What is the probability it is (a) red (b) yellow (c) green?
    Later, when half the sweets have been eaten, the probabilities of picking a red, a yellow or a green are 0, $\frac{1}{5}$ and $\frac{4}{5}$ respectively. How many of each colour have been eaten?
13. A spinner has the four numbers 1, 2, 3 and 4. It is spun twice and the scores are added together. Copy and complete the table for the sum of the two scores. State the probability of a sum of (a) 8 (b) 6 (c) 5 (d) 9 or more (e) less than 9.

	1	2	3	4
1				
2			5	6
3	4	5		
4				

14. Two dice are thrown together and the scores are added together. Make a table as in question **11**. State the probability of (a) 12 (b) 10 (c) 7 (d) 1 (e) an odd number.

15. A bag contains 1 green and 3 red marbles. A second bag contains 1 red and 3 green marbles. A marble is drawn from each bag. Copy and complete the table showing the possible pairs which can be obtained. State the probability of (a) two red (b) two green (c) one of each.

	R	G	G	G
R	RR		RG	
R		RG		
R				
G				GG

Index numbers

A certain article cost £20 in 1969 and £45 in 1979. The 1979 price can be expressed as a percentage of the 1969 price.

$$\frac{\text{1979 price}}{\text{1969 price}} = \frac{£45}{£20} = \frac{45}{20} \times \frac{100}{1}\% = 225\%$$

If we leave out the percentage sign, we have an index number. The 1969 price is called the base of the index number.

We can write

225 is the index number for the 1979 price (1969 = 100).

Suppose that the index number for the 1959 price is 60 (1969 = 100).

$$\text{Then } \frac{\text{1959 price}}{\text{1969 price}} = 60\%, \frac{\text{1959 price}}{£20} = \frac{60}{100}, \text{ 1959 price} = £12$$

The results are shown in the table:

Year	1959	1969	1979
Price	£12	£20	£45
Index Number (1969 = 100)	60	100	225
Index Number (1959 = 100)	100	y	x

Change of Base. If we use the 1959 price as base, the 1959 index number is 100. Let the index numbers for 1979 and 1969 be x and y.

$$\text{Then } \frac{x}{225} = \frac{100}{60} \text{ giving } x = 375 \text{ and } \frac{y}{100} = \frac{100}{60} \text{ giving } y = 167 \text{ approx.}$$

Exercise 41

1. An article cost £8 in 1970 and £14 in 1975. Calculate the index number for the 1975 price using the 1970 price as base. The same article cost £5.60 in 1965. Calculate the index for 1965 (1970 = 100). Show the results in a table as above.
2. | Year | 1972 | 1973 | 1974 |
   |---|---|---|---|
   | Price | £1.50 | £1.80 | £2.40 |
   | Index No. (1972 = 100) | 100 | k | n |

   Calculate k and n.
3. For the data of question 2, calculate index numbers for 1972 and 1973 using 1974 as base.
4. | Year | 1967 | 1972 | 1977 |
   |---|---|---|---|
   | Output (million tonnes) | 165 | 127 | 121 |
   | Index Number | 100 | g | h |

   The table shows the coal production for the United Kingdom in three years. Calculate g and h, correct to the nearest unit.
5. Use the data of question 4 to find index numbers for 1967 and 1972 with the 1977 output as base.
6. | Year | I | II | III |
   |---|---|---|---|
   | Index No. | 100 | 120 | 135 |
   | Price | £60 | a | b |

   Find the prices a and b.
7. | Year | 1976 | 1977 | 1978 |
   |---|---|---|---|
   | Index Number | 100 | 85 | 65 |
   | Price | £14 | c | d |

   The table is for a certain electronic calculator. Find the prices c and d.
8. (a) The prices of a standard loaf of bread at the ends of certain years were: 1967, 8p; 1972, 11p; 1977, $26\frac{1}{2}$p. Express this data in index form (1967 = 100).
   (b) The prices of a gallon of petrol for the same years were: 28p, 36p, $76\frac{1}{2}$p. Express this data in index form (1967 = 100), correct to the nearest unit.
   (c) Compare the results of (a) and (b).
9. The natural gas used in the United Kingdom in 1969 was equivalent to 9.2 million tons of coal. The figures for 1971, 1975 and 1977 were 28.4, 54.5 and 61.3. Express this data in index form (1969 = 100), correct to the nearest unit.
10. The indices of average earnings in Great Britain for various years are shown in the table, taking two different bases.

Year	1968	1970	1972	1974	1976
Index A	89	100	x	148	215
Index B	y	68	79	100	z

    Calculate x, y and z, correct to the nearest unit.

11. The table shows the index of house values for each of three years.

Year	I	II	III
Index	100	120	135

(a) A house was valued at £12 000 in Year I. What is its value in Year II?
(b) Another house was valued at £20 000 in Year I. What was its value in Year III?
(c) A third house was valued at £18 000 in Year II. What was its value in Year III?
(d) Taking Year II as base year, calculate the index number for Year III and for Year I.

Flow charts

Exercise 42

1.

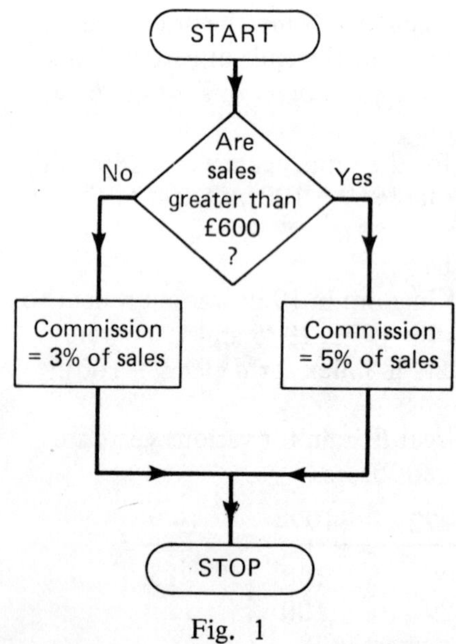

Fig. 1

Fig. 1 shows how the commission for a sales representative is calculated from his sales for a week. Use the chart to calculate the commission when the sales are (a) £400 (b) £800 (c) £968

2.

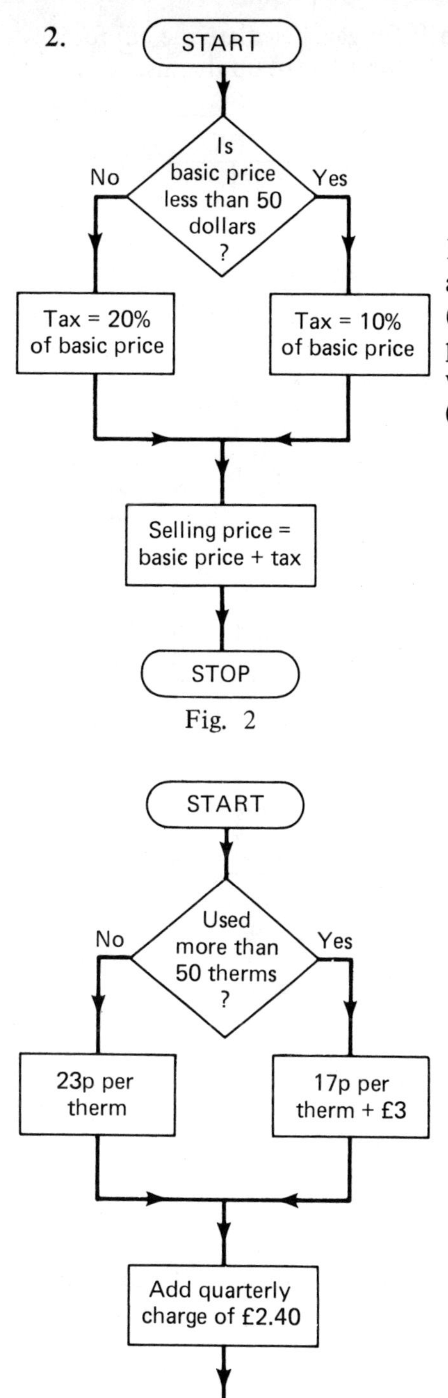

Fig. 2

Fig.3

In a certain country there is a tax on all articles sold in shops. Use the chart (Fig. 2) to calculate the price to be paid (after adding tax) for an article with a basic price of (a) 70 dollars (b) 30 dollars (c) 165 dollars.

3. Fig. 3 is a chart for a certain tarriff for gas. Calculate the bill for a quarter when the gas used is (a) 40 therms (b) 60 therms.

4. Fig. 4 is a chart for calculating the fee for parking in a shoppers' car park in a certain city. Calculate the fee for parking (a) for 40 minutes (b) for $2\frac{1}{2}$ hours (c) for 25 minutes (d) from 14 10 to 15 32.

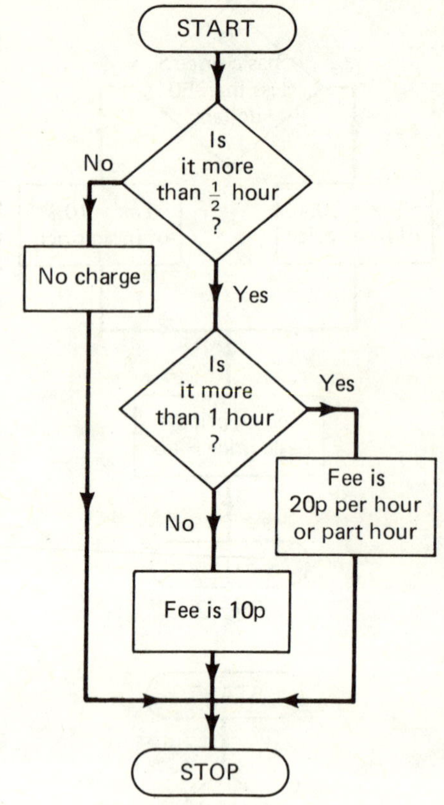

Fig.4

5. At sale time a shop gives a discount of 20% on the price of articles marked at less than £10 and a discount of 30% on the price of articles marked at £10 or more than £10. Draw a flow chart for finding sale prices from marked prices. Use your chart for an article marked at (a) £40 (b) £5 (c) £8. (The chart should be like Fig. 2.)

6. A mail order firm charges 50p for postage and packing for articles valued at £5 or less and 75p for articles valued at more than £5 but not more than £10. For articles valued at more than £10 there is no charge. Draw a flow chart for finding the postage and packing charges. (It should be like Fig. 4.)
Use your chart to find the postage and packing on articles valued at (a) £3.20 (b) £15.80 (c) £8.65.

Test Papers

Mathematical tables, slide rules and electronic calculators should not be used.

Test Paper 1

1. Evaluate: (a) $13 + 2(12 - 7)$ (b) $6.5 + 3.7 - 4.47$ (c) 0.07×0.4 (d) 0.063×200 (e) $50.4 \div 30$

2. Express as a single fraction:
   (a) $\frac{1}{4} + \frac{5}{6} - \frac{2}{3}$ (b) $1\frac{7}{8} \times \frac{2}{5}$ (c) $\frac{2}{1.5} \div \frac{7}{9}$

3. (a) Express 32.168 correct to 3 significant figures.
   (b) Express 0.4729 correct to 2 decimal places.
   (c) Express $\frac{5}{8}$ as a percentage.
   (d) Express 36% as a fraction in its lowest terms.

4. (a) An aircraft travels 45 kilometres in 3 minutes. How far does it go in 8 minutes at the same speed?
   (b) A tractor has front wheels of diameter 40 cm and back wheels of diameter 60 cm. When the front wheels turn 90 times, how many times do the back wheels turn?

5. A train left Exton at 22 10 h and arrived at Wyeford at 02 40 h. How long did the journey take? If the distance is 360 km, find the average speed of the train.

6. During a closing down sale the prices in a furniture shop were reduced by 30%. Find the sale prices of a chair originally priced at £60 and of a bookcase originally priced at £85.

7. In a certain district the local rate was fixed at 72 p in the £.
   (a) Calculate the amount to be paid in rates for a shop assessed at £550.
   (b) Calculate the rateable value of a house for which the rates were £126.

8. In a test, ten pupils obtained the following marks:
   6, 9, 4, 7, 8, 5, 6, 7, 7, 5.
   (a) Arrange the marks in order of size and state the median mark.
   (b) Calculate the mean mark.

Test Paper 2

1. (a) Calculate the squares of 0.3, 0.7 and $1\frac{1}{3}$.
(b) Find the square roots of 0.25, 0.04 and $2\frac{1}{4}$.
(c) Express 1225 as the product of prime numbers and hence find $\sqrt{1225}$ and $\sqrt{12.25}$.

2. Evaluate 7.9×0.86 and $25.97 \div 4.9$.

3. (a) Write out the first nine numbers in base three.
(b) Evaluate in base three $1 + 2$, $2 + 2$ and $100 - 1$.

4. The scale of a map is 1 cm to 50 km. On the map two towns are 8 cm apart. What is the real distance between them?
The real distance between two other towns is 480 km. What is the distance between them on the map?

5. (a) A rectangular field is 450 metres by 320 metres. Calculate its perimeter in kilometres and its area in hectares. (1 hectare = 10 000 m^2.)
(b) The area of a triangle is 35 cm^2. Its base is 14 cm. Calculate its height.

6. A discount of 15p in the £ was allowed on all marked prices during a sale. Find the marked prices of goods for which the sale prices were £2.55 and £13.60.

7. A house is insured for £28 000 and its contents for £6000. The rates are 14.5p per £100 for the buildings and 32p per £100 for the contents. Calculate the total premium to be paid.

8. (a) An ordinary die is thrown. State the probability of
(i) a six (ii) an even number (iii) a number less than 3.
(b) Two such dice are thrown. Find the probability of
(i) two sixes (ii) two even numbers.

Test Paper 3

1. Simplify: (a) $\frac{2}{5} + \frac{1}{2} - \frac{3}{4}$ (b) $2\frac{1}{3} \times 1\frac{5}{7}$ (c) $4\frac{2}{3} \div 1\frac{3}{4}$

2. (a) Calculate 7% of £28.
   (b) Express $\frac{5}{12}$ as a percentage correct to one decimal place.

3. Evaluate: (a) $20 - 4(10 - 7)$ (b) $11.3 + 8.8 - 1.62$ (c) $(0.08)^2$
   (d) $16.8 \div 700$ (e) $2.727 \div 0.9$

4. Express in the form $a \times 10^n$, where $1 \leqslant a < 10$ and n is an integer:
   (a) 814 (b) 0.0035 (c) 28×10^3 (d) $(3.2 \times 10^3) \times (4 \times 10^2)$.

5. (a) Express as base ten numbers 23_{four}, 23_{six} and 23_{eight}.
   (b) Express 23_{ten} in (i) base four (ii) base six (iii) base eight.

6. A cylindrical tin has a diameter of 7 cm and a height of 10 cm. Using $3\frac{1}{7}$ for π, calculate (a) the area of a label to cover the curved surface of the tin and (b) the volume of the tin.

7. A pupil with a school party in France cashed travellers' cheques for £20 into francs at the rate of £1 = 8.65 francs. He spent 124 francs and on arriving back at Dover he changed the remaining francs into British money at the rate of £1 = 8.80 francs. How much British money, to the nearest penny, did he receive?

8. The number of apples in each of 50 boxes was counted and the following table was obtained:

Number of apples per box	39	40	41	42	43	44	45
Frequency (number of boxes)	3	5	9	15	10	6	2

   (a) Draw a frequency polygon to represent this data.
   (b) Calculate the mean number of apples per box.

Test Paper 4

1. (a) Express the following ratios in their simplest forms:
14:21, 5:35, £3:£1.20, 600 m:2 km
(b) A sum of money was divided into three parts in the ratio 2:3:5. The largest part was £15. What were the other two parts?

2. (a) If VAT is levied at 15%, how much is added to a bill for £128?
(b) A portable TV set was sold for £69. Find the price before VAT was added.

3. Round each number to 1 significant figure and find the approximate value, correct to 1 significant figure, for the product or quotient.
(a) 3.2×1.8 (b) 7.1×8.9 (c) 98×32 (d) $0.057 \div 2.9$

4. (a) A rectangular tank has a base 45 cm by 40 cm. Calculate the area of the base. 27 000 cm^3 of water are poured into the tank. Calculate the height of the water.
(b) A circle has a radius of $10\frac{1}{2}$ cm. Using $\frac{22}{7}$ for π, calculate its circumference and its area.

5. (a) The length and breadth of a rectangle are 7.8 cm and 5.4 cm, each being correct to 1 decimal place. Which of the following gives the greatest possible area in cm^2?
(A) 7.9×5.3 (B) 7.7×5.5 (C) 7.9×5.5 (D) 7.75×5.35
(E) 7.85×5.45
(b) John estimates that he has run 52 m to the nearest metre in a time of 6 s to the nearest second. Which of the following gives the greatest possible value, in m/s, for his speed?
(A) $53 \div 7$ (B) $52.5 \div 6.5$ (C) $53 \div 5$ (D) $52.5 \div 5.5$ (E) $51.5 \div 5.5$

6. A washing machine can be purchased for £178 cash or a deposit of £30 and 24 monthly payments of £6.50. How much more is paid by the second method?

7. A householder used 760 units of electricity in a certain quarter. He had to pay a fixed charge of £3.10 and 2.95p for each unit used. Calculate his bill for the quarter.

8. The approximate areas of the four parts of the United Kingdom are as follows: England 13 million hectares, Wales 2 million hectares, Scotland $7\frac{2}{3}$ million hectares and Northern Ireland $1\frac{1}{3}$ million hectares. Draw a pie chart to represent this data.